T0326502

Sustainability
and Innovation

Sustainability and Innovation
The Next Global Industrial Revolution

Salah M. El-Haggar

with contributions by

Dina Abdel Alim

Mohamed Abu Khatwa

Hussein ElKersh

Irene Fahim

Rania Hamza

Lama El Hatow

Yasser Ibrahim

Mokhtar Kamel

Dalia A. Nakhla

The American University in Cairo Press
Cairo • New York

Exclusive distribution outside Egypt and North America by I.B.Tauris & Co Ltd., 6 Salem Road,
London, W2 4BU

Dar el Kutub No. 20024/13
ISBN 978 977 416 647 1

Dar el Kutub Cataloging-in-Publication Data

El-Haggar, Salah M.
        Sustainability and Innovation: The Next Global Industrial Revolution/ Salah M.
    El-Haggar.—Cairo: The American University in Cairo Press, 2015
        p.   cm.
        ISBN: 978 977 416 647 1
        Sustainable Development
        Ecology
        338.927

1 2 3 4 5   20 19 18 17 16

Designed by Sally Boylan
Printed in the United States of America

Contents

Foreword

Universities are society's stewards of the world's collective knowledge and wisdom, gathered, preserved, and enriched for future generations. Today, as we learn more about the role of human activity in irreversible change in our natural world, fulfilling our responsibility to the children and grandchildren of today's students only grows more critical.

The American University in Cairo's commitment to research, teaching, and outreach in sustainable development began even before the idea gained the global currency it has today. Since the creation of the Desert Development Center (now the Research Institute for Sustainable Environments, RISE) more than three decades ago, AUC has championed enhancing livelihoods through responsible innovation in agriculture, energy and water use, building design, and many other domains. Our new purpose-built campus in New Cairo was designed to be environmentally sensitive, from its orientation to the prevailing winds to its use of local stone and other building materials. Shortly after the establishment of its Office of Sustainability in September 2011, AUC published the first Carbon Footprint Report by a university in the Middle East and North Africa; the third edition was released in 2015.

Just as we challenge ourselves as an institution to measure our impact on our environment and manage it responsibly, we promote scientific research, public debate, and policy advocacy on sustainable development. In 2013, AUC launched Master of Science and professional diploma programs in sustainable development; both are the first of their kind in the region, and they are made possible by the strength and scope of our faculty research in sustainability.

Among the most distinguished of those faculty is the chair of our department of mechanical engineering, Dr. Salah El-Haggar, only a sampling of whose work is showcased in this volume. Long before it was widely recognized as the pressing issue that it is, Dr. El-Haggar was working on recycling and preservation of natural resources, closing the loop of production in domains as varied as plastics, consumer glass, and marble processing. Indeed, as he argues here, we need to look beyond sustainability as simply preservation; the solid wastes, effluents, and emissions we merely manage today will be the inputs—the not-so-raw materials—of the new industrial processes and products of tomorrow.

From building materials to fertilizers, from consumer products to cosmetics and nutritionally valuable foodstuffs, there are opportunities for innovation in industrial uses of recycled plastics, glass, stone, algae, and other apparently low-value by-products of human activity in the modern world. There are novel ways to manage recycling, replacing purchases of durable goods like cars, refrigerators, or computers with long-term leases, thereby ensuring retrieval and reuse of the component parts and materials. There are, in other words, ample supplies of clever, cost-effective, and creative ways to integrate the three vital pillars of sustainability—social justice, environmental responsibility, and economic viability—into the way we live so as to ensure that we fulfill our obligations to future generations.

With this volume, Salah El-Haggar challenges all of us to take up this assignment with his customary energy, wit, and inventiveness. We at AUC are very grateful for this challenge and intend to do our part, and we hope that many others are similarly inspired by this work.

Lisa Anderson
President
The American University in Cairo

Acknowledgments

I would like to thank everyone at the American University in Cairo Press who was involved in the publication of this book. Without their help, devotion, and dedicated efforts, this book would not have come to fruition. I would also like to thank Nigel Fletcher-Jones, the director of the AUC Press, for encouraging me to publish this book, for his rigorous stewardship, and for his valuable contributions. I would also like to acknowledge Johanna Baboukis and Noha Mohammed for their valuable comments during the editing process.

My sincere appreciation goes to all my students at the American University in Cairo, who have always been an integral part of my research projects and who provided substantial assistance in the preparation of this book. Special thanks go to Dina Abdel Alim, Mokhtar Kamel, Dalia Nakhla, Irene Fahim, Rania Hamza, Hussein ElKersh, Lama El-Hatow, Yasser Ibrahim, Mohamed Abu Khatwa, Mahmoud El Haggar, Mohamed Allam, Ossama Manaa, Islam Ramzy, Amr El Shafei, Mohamed El-Haggar, Mohamed Serag, and many other students.

Finally, I would like to thank my wife, Sadika, for her trusted advice over the past thirty-five years. Our shared dream of creating a more peaceful and sustainable world has been the inspiration that guided me to this journey toward sustainability.

Contributors

Dina Abdel Alim holds a bachelor's degree in mechanical engineering with a concentration in industrial and design engineering, and a master's degree in materials and environmental sciences, both from the American University in Cairo. She has worked in the field of enterprise resources planning as a junior supply chain and manufacturing consultant.

Mohamed Abou Khatwa is a research scientist with the Illinois Applied Research Institute. He earned his PhD in materials engineering from McMaster University in Hamilton, Ontario, Canada in 2009. His thesis examined recrystallization kinetics in AA6111 aluminum alloys. He earned his BSc in 1999 and his MSc in 2003 in mechanical engineering from the American University in Cairo.

Hussein ElKersh holds a master's degree in mechanical engineering from the American University in Cairo and a bachelor's degree in mechanical power and energy from Minia University. He received a graduate merit fellowship award in 2012 and a certificate of academic honor for outstanding academic achievement in the mechanical engineering graduate program in 2013. His thesis research concentrated on recycling lead crystal glass sludge to produce foam glass with a wide range of properties. He is currently working as a directional drilling engineer with Schlumberger.

Irene Fahim is an adjunct faculty member in the Department of Mechanical Engineering, American University in Cairo. Her PhD thesis project is developing novel polymer composite membranes for industrial applications. She earned her Masters of Material Science degree in mechanical engineering at AUC. Her thesis project was developing natural fiber (rice straw) reinforced composites. Fahim has experience in specialized software including Design Expert for optimization of experiment parameters, Minitab for statistics, Image J analysis software for image analysis, and Abacus for finite elements analysis.

Rania Hamza has been a PhD student in the Department of Civil Engineering at the University of Calgary (Canada) since 2014. Before starting her PhD studies, she worked as an engineer for the fluids, hydraulics, and construction surveying labs at the American University in Cairo. Hamza has more than ten years of experience in both academia and industry in the fields of environmental engineering and sustainable development. She received her MSc in environmental engineering (2001) and BSc in construction engineering (2006) from AUC. Her research interests include solid waste management, industrial solid waste recycling, and biological wastewater treatment. Her doctoral thesis focuses on treatment of high-strength industrial wastewater in an integrated anaerobic–aerobic granular sludge bioreactor.

Lama El Hatow is an environmental specialist focused on issues related to climate change and water resources. She holds an M.Sc. in environmental engineering from the American University in Cairo and is currently pursuing a PhD at Erasmus University Rotterdam in the Netherlands. She is the co-founder of the Water Institute of the Nile (WIN), a think tank for the basin-wide management of water resources of the Nile River. El Hatow works as an environmental consultant for a number of international organizations. She has been engaged in the UNFCCC international climate change negotiations for the last five years, focusing on climate policy.

Yasser Mohamed Ibrahim is the sector manager for environment and social integrity at the Multilateral Investment Guarantee Agency (MIGA), the political risk insurance and credit enhancement arm of the World Bank Group. He previously worked in Egypt as a project manager for Barclays Bank; as head of environment, health, and safety for

TIPOCO oil and gas services company; and as technical director for Global Environment Ltd. Ibrahim holds a master's degree in environmental engineering and a bachelor's degree in mechanical engineering from the American University in Cairo.

Mokhtar Kamel holds a bachelor's degree in industrial and systems engineering, and an MS in industrial engineering from the American University in Cairo, with a thesis topic of industrial and environmental engineering. He has worked as a researcher for UNESCO and as a technical translator, and is now a faculty member at Suez Canal University.

Dalia A. Nakhla holds a PhD in environmental engineering from the School of Science and Engineering, the American University in Cairo. Nakhla has a B.Sc. in construction engineering and a M.Sc. in environmental engineering. Nakhla has more than fifteen years of experience in the fields of environmental management, environmental impact assessment, and solid waste management.

CHAPTER 1

Sustainability and the Green Economy

Salah M. El-Haggar

"The earth offers enough for everyone's need, not for everyone's greed."
Mahatma Gandhi

Introduction

Sustainable development is a dynamic process that enables people to improve their quality of living in ways which simultaneously protect and enhance the earth's life support systems and preserve these resources for generations to come. In essence, sustainable development is about five key principles: quality of life, fairness and equity, participation and partnership, care for our environment, and respect for ecological constraints (Ben-Eli, 2009).

In 1987, the United Nations–sponsored Brundtland Commission released *Our Common Future* (WCED, 1987), a report that highlighted widespread concerns about the environment and poverty in many parts of the world. The Brundtland report noted that while economic development cannot stop, it must change course to fit within the planet's ecological limits. It also popularized the term 'sustainable development' (SD), which the report defines as "development that meets present needs without compromising the ability of future generations to meet their own needs." Sustainability includes economic, environmental, and social development. SD is not defined as "a fixed state of harmony, but rather a process of change in which the exploitation of resources, direction of investments, orientation of technological development, and institutional change are made consistent with future as well as present needs" (WCED, 1987). In engineering, the term 'sustainability' focuses primarily on the process of using energy and resources at a rate that does not dramatically affect the performance of the natural environment and the demands of future generations.

SD has become more of an issue in recent years due to depletion of both renewable and non-renewable resources, increases in human population, and problems such as climate change, deforestation, desertification, and species loss that they are causing. Therefore, any new or existing project has to be studied carefully to ensure its sustainability.

As can be seen, the definitions for the concept of sustainability seem to be very broad and not necessarily specific enough for everyone, or even professionals in the industry, to agree upon. As a result, it is important to understand how the word was initially derived, and then see how it developed. The word 'sustainability' is derived from the Latin word *sustinere*. The word *tenere* means 'to hold,' and the prefix *sus-* is a variant of *sub-*, meaning 'under' or 'below.' 'Sustain' thus means 'to support from below.' This simple dissection allows us to understand the idea of the term rather than the literal meaning, and that is to care, to protect, to maintain, to support, and so on.

The four pillars of SD are the technological, social, economic, and environmental aspects, which, if satisfied, may help to ensure sustainability. To help this become reality, a number of changes, such as new paradigms, values, visions, policies, education and training, indicators, facilitators, and formulas, will be needed to make the societal journey toward sustainability.

Cradle-to-Cradle System

Life cycle assessment (LCA) is a methodology for examining the environmental impacts associated with any product, from the initial gathering of raw materials from the earth until the point at which all residuals are returned to the earth ('disposal'). This is known as the "cradle-to-grave" (C2G) approach (ISO 14040:2006). Unfortunately, most manufacturing processes since the industrial revolution have been based on this one-way, C2G flow of materials. The C2G flow of materials has proven to be inefficient because it depletes natural resources. The "cradle-to-cradle" (C2C) concept promotes SD in a wider approach. It is a system of thinking based on the belief that human endeavors can emulate nature's elegant system of safe and regenerative productivity by transforming industries to sustainable enterprises and eliminating the concept of waste (El Haggar, 2007).

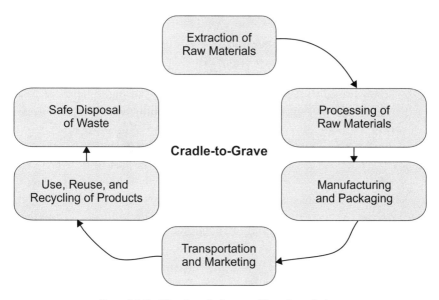

Figure 1.1. Traditional cradle-to-grave life cycle analysis

Environmental protection is very important for SD as well as for conservation of natural resources. In other words, transforming waste or emissions into by-products or co-products is a must for sustainability. Unsustainable human activities are creating an open-loop cradle-to-grave cycle (fig. 1.1) that cannot continue and has to be removed from the conceptual and operational framework of our societies. Closing the loop for renewable resources and producing zero waste can be managed by changing from the cradle-to-grave (C2G) system, where waste is disposed of, to a cradle-to-cradle (C2C) system, where waste is recycled into usable resources, as will be discussed in detail in chapter 2.

The C2C system depicted in figure 1.2 is one where item production is changed from a one-way process into a cyclic system. This method helps to ensure that the products, after use, are returned to the original manufacturer or to other manufacturers to reprocess them into a new product, thereby reducing the amounts of raw materials needed for production. A proper life cycle assessment (LCA) must be conducted to decide how they will be managed throughout all phases of their cycle: product design, materials and energy acquisition prior to production, transformation into products, usage of wasted resources from production in other processes, the consumer usage phase, and the transformation of the 'dead' product into other products for subsequent cycles. All of this can make contributions toward societal SD. But energy is needed at every step and it is never possible to transform 100% of the materials into other products. Furthermore, there is always some degree of contamination and a consequent downgrading of the quality of the materials being recycled. But, through innovations, downcycling can be converted to upcycling, as will be discussed throughout the book.

A new hierarchy for waste management reflecting the C2C concept was developed at the American University in Cairo in 2001 and upgraded in 2003. Named the "7-R rule" or "7-R cradle-to-cradle approach," the concept starts from developing *regulations* for *reduction* at the source, *reuse, recycle, recovery* by sustainable treatment for possible material recovery (rather than waste-to-energy recovery). The last two Rs are *rethinking* and *re-innovation*, in which people should rethink their waste (qualitatively and quantitatively) before taking action for treatment or disposal and develop a renovative/innovative technique to solve the waste problem (El-Haggar, 2007). The 7-R approach is based on the concept of adopting the best practicable environmental option (from

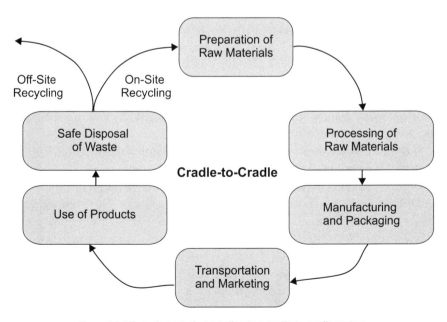

Figure 1.2. Life cycle analysis according to a cradle-to-cradle system

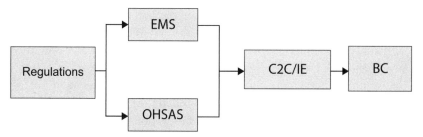

Figure 1.3. Proposed sustainable development road map

not only the technical but also the economic and social point of view) for individual waste streams and dealing with waste as a by-product. But while the approach can help society make some progress on its journey toward a more sustainable society, it cannot do so on its own. So what else is needed? Certainly major paradigm shifts, new visions, values, ethical standards, climate neutrality based upon renewable resources, renewable energy, improved energy efficiency, sustainable stewardship ethical standards, and so on may help societies to make further progress. This 7-R rule will ban disposal and treatment facilities and transform what was previously wasted into new products in a much more responsible management of renewable and non-renewable resources.

Natural resources are a crucial issue for SD because finding new sources of raw material is becoming costly and difficult. Thousands of species are going extinct due to ignorance, greed, and global human overpopulation far beyond the ecological carrying capacity of the earth. Concurrently, the cost of treatment and safe management of waste is increasing exponentially and locating waste disposal sites is becoming more and more difficult. The impact of waste disposal on the environment is significant since it can contaminate air, soil, and water. To make waste management more sustainable, it should be moved from the traditional LSA following the C2G system to a new system, not relying on disposal facilities, to become an integrated and multi-life-cycle C2C system.

Proposed Sustainable Development Road Map

The proposed SD road map is shown in figure 1.3. The elements are regulations; environmental management systems (EMS); cleaner production (CP); occupational health and safety (OHS); industrial ecology (IE) according to C2C; and moving beyond compliance (BC) with regulations. The initial procedure in the proposed road map is developing a set of environmental regulations, the strict enforcement of which would force investors to implement the EMS (ISO 14001) within the organization's policy and decision-making strategies so as to identify waste and pollution. The information gathered from the EMS analysis determines whether the goals have been met or not. The main goal is not only compliance with regulations but BC for conservation of natural resources, which will finally lead to SD.

Regulations

Regulations are a basis for achieving SD. These regulations are set by governments to provide organizations and projects with the policies they must abide by to reach sustainability. The regulations provide for fines and other incentive mechanisms to encourage 100% sustainability by providing monetary rewards to those who meet sustainability requirements, funded by fines collected from those who fail to meet them.

Beyond Compliance

Beyond Compliance (BC) is a new concept whereby an organization aims not merely for compliance with regulations or requirements but to target efficiency. BC is a shift in perspective toward pursuing sustainability, a shift from a reactive attitude to proactive innovation. The paradigm shift will start with compliance. Compliance with the regulations does not mean that a particular industry has achieved sustainability; it is only the first step in the journey.

Environmental Management Systems (EMS) and Cleaner Production (CP)

One of the main elements of SD is integrating environmental management systems (EMS) or ISO 14001 within the day-to-day activity of an operation. An EMS consists of a systematic process that allows an organization to assess, manage, and reduce environmental hazards. The continuous monitoring of the environmental impacts of an organization's activities is integrated into the actual management system, guaranteeing its continuation as well as commitment to its success.

The EMS is a part of the overall management system of an organization, which consists of organizational structure, planning, activities, responsibilities, practices, procedures, process, and resources for developing, implementing, achieving, reviewing, and maintaining the environmental policy (El-Haggar and Sakr, 2006).

Continuous development and implementation of an EMS provides a number of benefits to a company:
- financial benefits (through cost savings as well as increased competitiveness in local and international markets)
- improved performance and image for the company
- reduced business risks
- compliance with environmental regulations

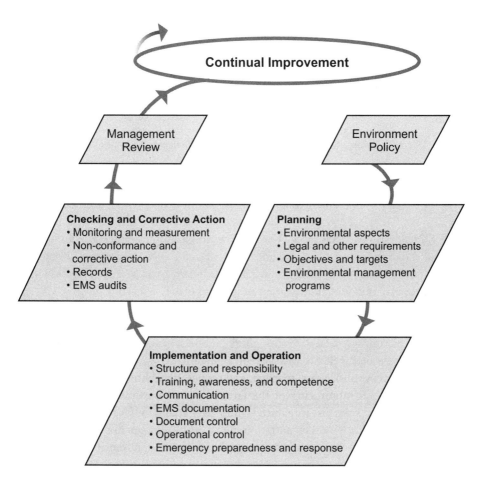

Figure 1.4. EMS model (El-Haggar, 2007)

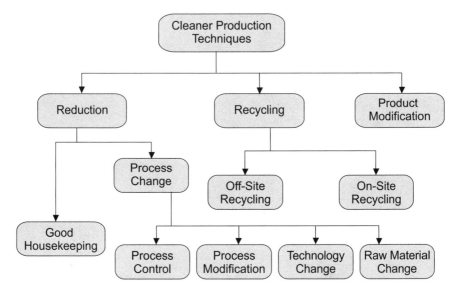

Figure 1.5. Cleaner Production techniques (El-Haggar, 2007)

Periodic EMS audits must be carried out to check that the EMS is effectively implemented and maintained. An EMS is also a necessary tool for Cleaner Production (CP), which focuses on the prevention of waste generation at the source. This is achieved by adopting the CP techniques shown in figure 1.5 to enhance processes, products, or services that will lead to savings in energy, raw material, and costs, as well as protection of the environment and natural resources in order to reach C2C.

EMS integrated with CP are the primary SD tools. Major efforts are made in applying these concepts worldwide, especially in developing countries, because of the immediate environmental and financial benefits they generate if properly applied.

Applying EMS across existing activities, in addition to adopting the 7-R Rule, guarantees the organization will be in compliance with environmental regulations, which will in turn facilitate the implementation of CP techniques. The adoption of the 7-R Rule does not rely solely on investors; research institutes and universities should develop solutions to existing environmental problems and promote the concept of SD. At the same time, new investors should be encouraged to cooperate and establish a recycling unit to reuse and recycle waste and produce raw materials and products that can be sold.

EMS can be implemented using CP techniques or pollution control systems. The key difference between CP and other methods like pollution control is the choice of the timing, cost, and sustainability. Pollution control follows a 'react and treat' rule, while cleaner production adopts a 'prevent rather than cure' approach (El-Haggar, 2007). Thus, CP focuses on before-the-event techniques, as displayed in figure 1.5:

1. Source reduction
- Good housekeeping
- Better product design
- Better product-service system design
- Process changes
    - Better process control
    - Equipment modification
    - Technology change
    - Input material change

2. Recycling
- On-site recycling
- Useful by-products through off-site recycling

3. Product modification

CP can reduce operating costs, improve profitability and worker safety, and reduce the environmental impact of the business. Companies are frequently surprised at the cost reductions achievable through the adoption of CP techniques. Frequently, minimal or no capital expenditure is required to achieve worthwhile gains, with fast payback periods. CP techniques benefit industry in a number of ways (El-Haggar, 2007):
- Reduced waste disposal costs
- Reduced raw material costs
- Reduced health–safety–environment (HSE) damage costs
- Improved public image and public relations
- Improved company performance
- Improved competitiveness in local and international markets
- Improved compliance with environmental protection regulations
- Reduced toxic substance usage and toxic substance risks to humans and the environment
- Reduced energy usage throughout product life cycles

On a broader scale, CP can help alleviate the serious and increasing problems of air and water pollution, ozone depletion, global warming, landscape degradation, solid and liquid wastes, resource depletion, acidification of the natural and built environment, visual pollution, and reduced biodiversity.

The EMS can provide a company with a decision-support system and action plan to bring CP into the company's strategy, management, and day-to-day operations. As a result, EMS will provide a tool for CP implementation and pave the road toward it. Therefore, integrating CP techniques with EMS will help the system to approach zero pollution (C2C), promote compliance with environmental regulations (El-Haggar and Sakr, 2006), reach BC, and maximize the benefits when CP and EMS benefits are integrated into the same system.

CP can be incorporated into the environmental policy of the organization as a commitment from the top management to encourage the organization to seek CP techniques everywhere as a solution to any environmental problem. During the planning phase of EMS, CP should be the main tool to achieve the objectives and targets.

Occupational health and safety

The Occupational Health and Safety Assessment Series (BS OHSAS 18001:2007) has been developed to minimize occupational health and safety (OHS) risks. OHSAS 18001 is an international OHS management system specification designed to enable companies to eliminate or minimize the risk to employees and others who may be exposed to OHS risks associated with the business's activities. It also allows companies to demonstrate their commitment to providing a safe working environment, protecting their employees, and improving overall performance.

Because of the success of the ISO 14001 standards for environmental management, OHSAS 18001 was developed to be compatible with other ISO management systems. Many sections and sub-clauses are similar, such as management review, document control, and corrective and preventive actions, making OHSAS 18001 easily compatible with ISO 14001 certifications.

OHSAS 18001 focuses on the identification, elimination, and continual improvement of hazards and risks within the work environment. The methodology of managing OHSAS is based on planning for hazard identification, risk assessment, and risk control. The benefits of implementing OHSAS 18001 include:

- Reduced injuries
- Reduced insurance costs and liability
- Reduced costs due to personal injury and sick leaves
- Reduced human resource constraints from personal injuries
- Ease of managing safety risks
- Ease of managing legal and compliance requirements
- Enhanced employee safety awareness
- Enhanced public image

The application of EMS and registration of ISO 14001 helps an organization implement ISO 18001 with continual performance improvement, as shown in figure 1.4.

Life cycle assessment/cradle-to-cradle

In the proposed framework structure for SD shown in figure 1.3, the current LCA will be modified to evaluate industrial activities adopting C2C concepts. Ideally, there should either be no impact on the environment or it will be within the allowable limits. The lower the impact, the more efficiently the C2C concept is adopted, and the closer the organization is to resembling a natural ecosystem, thereby complying with regulations and realizing SD.

Industrial ecology (IE)

IE has recently enjoyed increased interest as both an object of academic study and a practical policy tool (Ehrenfeld, 2004). IE can be regarded as a science of sustainability. It offers great promise for improving the efficiency of human use of the ecosystem. Technological improvements cannot be fully sustainable without taking the environment into consideration; zero pollution or C2C is a must for IE. Cooperation and community are also important components of the ecological metaphor of sustainability.

The main tool for IE is CP, which provides a framework for identifying the impact of industries on the environment and for implementing strategies to reduce this impact, as it involves studying the interactions and relations between industrial and ecological systems. The ultimate goal of IE is to achieve SD that will eventually lead to compliance with environmental regulations, protecting the environment and conserving natural resources.

IE aims at transforming industries to resemble natural ecosystems where any available source of material or energy is consumed by some organism. The managerial approach to IE involves analyzing the interaction between industry and the environment, through the use of tools such as LCA. "In recent years, concepts drawn from industrial ecology have been used to plan and develop eco-industrial parks [EIPs] that seek to increase business competitiveness, reduce waste and pollution, create jobs, and improve working conditions" (Gibbs and Deutz, 2005). The technical approach, on the other hand, involves implementing new process and product design techniques such as CP and EIPs. The interaction of CP techniques and LCA finally leads to IE. A set of design rules that promote IE includes:

- Closing material loops
- Using energy in a thermodynamically efficient manner, employing energy and material cascades
- Avoiding upsetting the system's metabolism
- Eliminating materials or wastes that upset living or inanimate components of the system
- Dematerializing, or delivering the function with fewer materials (Ehrenfeld, 1997)

Basic Tools of Sustainable Development

IE requires an indicator to evaluate the performance of industries in implementing IE tools. C2C is the main indicator of IE. The degree to which C2C is achieved will give an indication of how closely an industry is emulating nature's ecosystems. Achieving IE will lead to sustainability.

Hidden Tools of Sustainable Development

The hidden (or indirect) tools for SD, to be included at all times throughout the SD road map, are ethical investment and sustainability reporting. Ethical investment (EI) is an investment strategy that seeks to maximize both financial return and social good. EI is also known as socially responsible investment (O'Rourke, 2003). It favors corporate practices that promote environmental stewardship, consumer protection, human rights, and diversity. They might avoid businesses involved in alcohol, tobacco, gambling, weapons, abortion, and so on.

The US NGO called the Global Reporting Initiative (GRI) is a leading organization in the sustainability field, providing one of the world's most powerful standards for sustainability reporting. It was launched in 1997 as a joint initiative of GRI and the United Nations Environmental Programme. Sustainability reporting is a form of value reporting in which an organization publicly communicates its economic, environmental, and social performance. GRI promotes the idea of making sustainability reporting as common a practice as financial reporting. GRI guidelines apply to corporate businesses, public agencies, smaller enterprises, NGOs, and industry groups.

Reporting on sustainability performance is an important way for organizations to manage their impact on sustainable development. In addition, sustainability reporting promotes transparency and accountability. Reporting leads to improved sustainable development outcomes because it allows organizations to measure, track, and improve their performance on specific issues. Organizations are much more likely to manage effectively an issue that they can measure. The framework of reporting as suggested by GRI is:

- The action taken to improve economic, environmental, and social performance
- The outcomes of such action
- The future strategy for improvement

Sustainable Development Facilitators

Education for SD, environmental awareness, information technology, and social responsibility can be considered as facilitators for SD that can catalyze and facilitate the participatory process. They may help people and organizations to clarify their visions, ethical principles, perceptions, attitudes, and knowledge, so as to attain self-development and compliance as well as conservation of natural resources.

Education for Sustainable Development

"Education is the foundation for achieving sustainable development" (McCormick et al., 2005). "Education for Sustainable Development (ESD) is a learning process based on the ideals and principles that underlie sustainability. ESD supports five fundamental types of learning—learning to know, learning to be, learning to live together, learning to do, and learning to transform oneself and society" (Wals, 2009). The aim of ESD is to help people to develop attitudes, skills, and knowledge. ESD must be seen as a comprehensive package for quality education and learning within which key issues such as poverty reduction, environmental awareness, sustainable livelihoods, climate change, gender equality, corporate social responsibility, and protection of original cultures are found. The holistic nature of ESD allows it to be a possible tool for the achievement of Millennium Development Goals (MDGs) and

Education for All goals. Both of these initiatives have a set of objectives to be achieved by a certain deadline. ESD could be perceived as the vehicle for achieving those objectives. The MDG proposes to improve the health, nutrition, and well-being of some 1.2 billion humans who live on less than the equivalent of $1 a day (Nelson, 2007).

The United Nations Decade of Education for Sustainable Development (January 2005 to December 2014), for which UNESCO is the lead agency, seeks to integrate the principles, values, and practices of SD into all aspects of education and learning, in order to address the social, economic, cultural, and environmental problems we face in the twenty-first century (Wals, 2009). The contribution from the higher education sector for this initiative has been recognized internationally (Jones, Trier, and Richards, 2008) by bodies such as the Higher Education Funding Council for England (HEFCE), which promotes higher education as a major contributor to a society's sustainability.

Environmental awareness

The main catalyst of the proposed SD road map is awareness. Environmental awareness is an essential factor in each and every phase of the framework. Awareness will motivate individuals—especially investors—to carry out the various tasks and draw their attention to the benefits of preserving the environment, and the damage that could otherwise take place in the long run. Eco-efficiency can lead to major economic gains and increased efficiency. In the short run, activities that harm the environment are more attractive, due to either their low costs or ease of implementation. However, in the long run the damage costs can be significant and irreversible. Environmental awareness can be enhanced through ESD and information technology tools and principles.

Information technology

Since the 1980s, computer-based information technology has become the focus of global information. The environmental situation is continually changing as a result of human activities, so it is very important to obtain accurate and timely information about various environmental changes. This process will also contribute to global knowledge about the environment (UNEP, 1997).

One of the main tools for information technology is the Internet. The Internet is a two-way medium to communicate, which can give as well as receive information on bottom-up environmental activities. It improves interaction between people and institutions, facilitates access to government by those governed, and eases access to information in general. Barriers not directly related to the technology itself, such as illiteracy, computer illiteracy, and differences in the technical and educational qualifications of users, can be overcome.

IT can help promote the concept of environmental awareness (EA), which might increase the public's interest in the environment. People's response to improving their environment depends on the depth of their perception of environmental problems, and their willingness to take action. Participatory EA, such as bottom-up approaches, helps them to prioritize problems, understand how to solve these problems, and encourage them to be involved in the planning and implementation stages. To make participatory EA a success, an efficient information process is required as a tool. IT is the tool that matches the needs of modern times, although it faces several challenges in developing countries.

Public EA will not be created overnight; it grows gradually and slowly. The government cannot take the major role in this process, which must be controlled and directed by the public. The required role of the government, NGOs, and educated individuals is to find efficient and creative means to reach out to the public through IT.

Social responsibility (SR)
Social responsibility, also known as corporate social responsibility (CSR), corporate responsibility, corporate citizenship, responsible business, sustainable responsible business, or corporate social performance, is a form of corporate self-regulation integrated into a business model (Cramer, 2008). The European Commission (EC, 2001) defines CSR as "a concept where companies integrate social and environmental concerns in their business operation and in their interaction with their stakeholders on a voluntary basis." CSR is usually referred to as the "triple bottom line principal" (Dyllick and Hockerts, 2002).

The need for an international standard for SR was first identified in 2001 by the International Organization for Standardization (ISO) Committee on Consumer Policy. In 2003, the ISO Ad Hoc Group on SR completed an extensive overview of SR initiatives. In 2004, the ISO held an international, multi-stakeholder conference to discuss the launching of SR work with its main stakeholder groups: industry, government, labor, consumers, NGOs, service, support, research, and others, as well as a geographical and gender-based balance of participants (http://www.iso.org/is/social responsibility.pdf). The positive recommendation of this conference led to the establishment in late 2004 of the ISO Working Group on Social Responsibility to develop the future ISO 26000 standards.

Based on the success of ISO standards for organization and product development, the ISO decided to develop an international standard providing guidelines for SR. This standard was named ISO 26000. It offers guidance on socially responsible behavior toward the community and possible actions; it does not contain requirements and, therefore, in contrast to ISO 14001, is not certifiable. The ISO 26000 cannot be used as a basis for audits, conformity tests, and certificates. It can be used as a guidance document, on a voluntary basis, and encourages organizations to discuss their SR issues and possible actions with relevant stakeholders.

The ISO 26000 can be considered one of the major facilitators not only for a company to achieve SD but also for the community to achieve SD. It is a win-win scenario for company and community in the journey toward SD. It applies to both the public and the private sectors and is essential for industry, governmental, and non-governmental organizations as well as end users.

Green Economy

A green/sustainable economy is an economic development model based on sustainable development and knowledge of ecological economics. Its objectives are improved human well-being, social equity, and reduced environmental risks and ecological scarcities (UNEP, 2013). Accelerating the transition to a cleaner and greener economy has become a high priority for all countries. But transforming the economies of nations to be more sustainable will not be accomplished without more innovation and cooperation among these nations. Also, citizens will need to change their business and personal practices to support developing new, creative, and innovative technologies to promote green economy. Multidisciplinary teams of architects, engineers, scientists, business managers, financial experts, lawyers, entrepreneurs, political leaders, and resource managers are required in order to achieve a green economy (Elder, 2009).

'Green economy,' 'green growth,' 'sustainable economy,' and 'qualitative growth' are used interchangeably to mean basically the same concept. There is no agreed-upon, consistent definition for 'green economy.' The most widely used is the one introduced by UNEP. It defines greening the economy as the "process of reconfiguring businesses and infrastructure to deliver better return on natural, human, and economic capital investments, while at the same time reducing greenhouse gas emissions, extracting and using less natural resources, creating less waste, and reducing social disparities" (UNEP, 2009). In a green economy, economic growth and increase of employment opportunities and income are achieved by the investments that reduce pollution and preserve the natural resources. The main drive behind the green economy trend is the global financial and economic crisis that has emerged in the last decade. According to the UNEP, one major reason for this crisis is the misallocation of investments where the majority of capital was used in non-renewable projects such as fossil fuels, whereas very little capital was invested in renewable projects such as sustainable transportation or renewable energy (UNEP, 2011). Additionally, the polluters do not pay the cost of pollution. These incorrect economic strategies have led to the depletion and degradation of natural capital.

Therefore, the very broad definition of a green economy is an economy that promotes sustainable development together with economic growth. Its many benefits include improving people's lives, and sustaining and advancing economic, environmental, and social well-being, as shown in figure 1.6. Thus, the green economy can be better defined as an economy that implements sustainable development

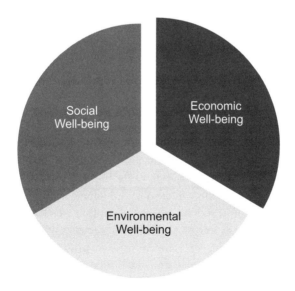

Figure 1.6. Definition of a green economy (Bapna and Talberth, 2011)

together with economic growth in order to approach social justice as well as poverty alleviation.

In past decades traditional economies have focused on generating wealth, with no concern that it might affect the society or the environment. A green economy, or green growth, is different from a traditional economy in that it is considered "more than a reordering of priorities; it involves a significant rethinking of the assumptions upon which the traditional economy has been based" (Halle, 2011). Therefore, a green economy is considered a win-win-win situation for the investor, society, and the environment.

The Global Citizens Center defines a green economy as (1) environmentally sustainable, (2) socially just, and (3) locally rooted. Environmental sustainability is based on the notion that the biosphere is a closed system with limited resources that are hard to regenerate, and so the economic systems should respect the sustainability of the ecosystem. As for social justice, culture and human dignity are considered as important as natural resources, and need to be preserved through ensuring an acceptable standard of living and equal opportunity for personal and social development. The third requirement, local rooting, is important for sustainability and justice and is achieved when communities encourage local production and exchange of goods and services ("Green Economy," 2013).

The main sectors of the green economy are energy, transportation, water, waste, and land management, in addition to green buildings (Burkart, 2013). The green economy can be subdivided into low-carbon economy, circular economy, sustainable consumption and production (SCP), and green growth.

The concept of a low-carbon economy has become more important with the rising public awareness of climate change and the urgent need for transformational change of the economy. In a low-carbon economy, the target is to emit a minimal amount of carbon dioxide and other greenhouse gases. Investing in renewable energy leads not only to new sources of income and jobs, but also to reducing carbon emissions.

A circular economy reduces the consumption of resources and the generation of different wastes. It reuses and recycles wastes through the production, circulation, and consumption processes. Investing in resource-efficient technologies and waste management/recycling generates new sources of income and jobs and also improves resource efficiency and waste management.

SCP tends to minimize the use of resources and the generation of wastes and pollutants over the full life cycle of products. Reduction, reuse, and recycling of resources in both production and consumption processes are also essential (Fulai, 2010).

The green growth component seeks to improve the quality and effectiveness of green growth planning and implementation by scaling up successful green innovations from the grassroots level to larger political initiatives. The World Bank and the Global Green Growth Institute are working together through the Green Growth Knowledge Platform to develop measurements for green growth. Indicators on green growth can help the public to better understand how it can contribute to and support the transition to a green economy. A solid base of statistical data will be needed to produce economic and environmental indicators that focus on different sectors such as energy, industry, water, and waste management. It is also possible to measure green economy by measuring the presence of green jobs, green markets, and investment in green industries (World Bank, 2013).

The green economy is a precious concept that can be treated as a vehicle to deliver three major outcomes: new sources of income and jobs; low-carbon emissions; and contributions to broader societal goals of sustainable development, social equity, and poverty reduction (Fulai, 2010). However, adopting green economy practices is not an easy task and requires a lot of innovation.

Case studies of green economy

The evolution to sustainable economy still has a long way to go; however, many countries are taking the lead by approving national green-growth or low-carbon economic policies. There are many successful examples of programs that help to increase productivity sustainably (WRI, 2013). For example:

- The Republic of Korea approved a national strategy and a five-year plan for green productivity for the period 2009–13. The program allocated 2% of the country's gross domestic product to investment in several green sectors such as renewable energy, energy efficiency, clean technology, and water.
- In Mexico City, investment in Bus Rapid Transit (BRT), a bus system that uses dedicated lanes on city streets, has decreased commuting times, reduced air pollution, and improved access to public transit for those who cannot afford private cars.

- China has the highest investment of any country in renewable energy. Its total installed wind capacity grew 64% in 2010. Its national policy sees clean energy as a major market in the near future.
- Namibia is handling its natural resources in such a way as to generate economic, social, and environmental benefits. The Namibian government provides economic incentives for sustainably-run food and employment in rural areas. More than half of the jobs are filled by women, and wildlife populations have increased.

"The principal challenge is how we move toward an economic system that will benefit more people over the long run" (WRI, 2013). The conversion to a green economy will require a fundamental change in thought and innovation of development, manufacture of goods and services, and customer habits.

The United Nations Development Programme (UNDP) supports an array of nationally defined development activities such as greening the value chain, which is applied by countries like Indonesia, the world's largest palm oil producer with earnings of $13.7 billion in 2008. Oil palm plantations covered 7.5 million hectares in 2010/11, up 300,000 hectares over the last decade. However, increasing palm oil production usually displaces local communities and indigenous peoples from their lands and livelihoods and impacts endangered species. The environment is affected as 3 million hectares of forested peat lands are converted to palm oil plantations, and deforestation is increased as 16 million hectares of tropical forests and peat lands are cleared for palm oil. The Green Commodities Facility initiative, supported by the UNDP, is helping the government target 7 million to 14 million hectares of degraded and abandoned lands that could be used to plant palm oil instead of clearing new forests and peat lands. This solution will reduce the environmental impact and promote sustainable livelihoods for small producers. Economic incentives to make sustainable palm oil profitable include improvements in productivity and yield, certification premiums, reducing emission from deforestation and forest degradation (REDD), carbon payments, and tax incentives (UNDP, 2013).

Promotion of green economy

To shift toward a green economy, traditional economic strategies and techniques should be changed. Economic mechanisms such as taxes, subsidies, trading schemes, financial reforms, and incentives could be used. Other regulatory schemes include setting policies and setting national regulations and standards such as the Green Funds scheme, a Dutch system of incentives developed to fund green projects (Le Blanc, 2011). Individual clients who invest their money in the banks at a lower interest rate get a tax rebate in return. At the same time, the banks offer cheaper loans to other investors who invest their money in green projects, hence encouraging investment in green projects. Another example is the subsidization of oil in many countries in the Middle East, which encourages the use of oil as an energy source and discourages the adoption of clean sources of energy. An alternative green economic practice is to subsidize the use of renewable and alternative energy sources. There have been many similar initiatives in Asia. For example, Vietnam was the first Asian country to adopt legislation that charges a tax on environmentally harmful substances.

On the national level, there should be a direction to invest in green projects to encourage innovation. This is what Japan's government did in the 1970s, becoming one of the primary world economies to invest highly in green research and development projects (UNEP, 2011).

Questions

1. Discuss the four pillars of SD.
2. Compare the benefits of EMS and CP and propose a method to integrate EMS with CP as a modification to ISO 14001.
3. Do you recommend modifying EMS/ISO 14001 to be inte grated with CP? Why? How can it be done?
4. Develop a CSR guideline for an international company to promote ESD.
5. What is the difference between direct tools and indirect tools in encouraging the implementation of SD?
6. Compare C2C and C2G and apply both concepts to municipal solid waste.
7. Develop a formula to quantify SD in an industrial activity.
8. Compare traditional economy and green economy, giving examples.
9. Develop a techno-financial mechanism(s) to promote green economy.
10. What is the difference between a green economy and a circular economy?

References

Bapna, Manish, and John Talberth. 2011. "Q&A: What is a 'Green Economy?' World Resources Institute. www.wri.org/stories/2011/04/qa-what-green-economy

Ben-Eli, Michael. 2009. "Sustainability: The Five Core Principles." The Sustainability Laboratories. www.sustainabilitylabs.org/page/sustainability-five-core-principles

BS OHSAS 18001:2007. Occupational Health and Safety Standards.

Burkart, K. 2013. "How Do You Define the 'Green' Economy?" www.mnn.com/green-tech/research-innovations/blogs/how-do-you-define-the-green-economy

Cramer, J.M. 2008. "Organising Corporate Social Responsibility in International Product Chains." *Journal of Cleaner Production* 16: 395–400.

Dyllick, T., and K. Hockerts. 2002. "Beyond the Business Case for Corporate Sustainability." *Business Strategy and the Environment* 11: 130–41.

EC (European Commission). 2001. "Promoting a European Framework for Corporate Social Responsibility." Green Paper. europa.eu.int/comm/employment_social/soc-dial/csr/csr2002_en.pdf

Ehrenfeld, S. 2004. "Industrial Ecology: A New Field or Only a Metaphor?" *Journal of Cleaner Production* 12: 825–31.

———. 1997. "Industrial Ecology: A New Framework for Product and Process Design." *Journal of Cleaner Production* 5, no. 1–2: 87–95.

Elder, J. 2009. "Fueling the Green Economy." *Community College Journal* 80, no. 2: 40–41.

Fulai, S. 2010. "A Green Economy: Conceptual Issues." Background paper for the UNEP, Major Groups and Stakeholders Consultation on Green Economy. Geneva, 26 October.

Gibbs, D., and P. Deutz. 2005. "Implementing Industrial Ecology? Planning for Eco-industrial Parks in the USA." *Geoforum* 36: 452–64.

"Green Economy." 2013. en.wikipedia.org/wiki/Green_economy#cite_note-5.

El-Haggar, S.M. 2007. *Sustainable Industrial Design and Waste Management: Cradle-to-Cradle for Sustainable Development*. Boston: Elsevier Academic Press.

El-Haggar, S.M., and D.A. Sakr. 2006. "Nezzam al-edara al-beaea wa-l-tekhnologiya: ISO14001^{Plus}." Cleaner Production Technologies Series 3. Cairo: Dar al-Fikr al-'Arabi.

Halle, M. 2001. "Accountability in the Green Economy." *Review of Policy Research* 28, no. 5.

ISO 14040:2006. *Environmental Management—Life Cycle Assessment—Principles and Framework*. Geneva: International Organization for Standardization.

ISO Standard. www.iso.org/iso/socialresponsibility.pdf

Jones, P., C.J. Trier, and J.P. Richards. 2008. "Embedding Education for Sustainable Development in Higher Education: A Case Study Examining Common Challenges and Opportunities for Undergraduate Programs." *International Journal of Educational Research* 47: 341–50.

Le Blanc, D. 2011. "Special Issue on Green Economy and Sustainable Development." Natural Resources Forum 35. www.ieg.earthsystemgovernance.org/sites/default/files/files/publications/Special%20issue%20on%20green%20economy%20and%20sustainable%20development%20Le%20Blanc.pdf

McCormick, K., E. Muhlhauser, B. Norden, L. Hansson, C. Foung, P. Arnfalk, M. Karlsson, and D. Pigretti. 2005. "Education for Sustainable Development and the Young Masters Program." *Journal of Cleaner Production* 13: 1107–12.

Nelson, P.J. 2007. "Human Rights, the Millennium Development Goals, and the Future of Development Cooperation." *World Development* 35, no. 12: 2041–55.

O'Rourke, A. 2003. "The Message and Methods of Ethical Investment." *Journal of Cleaner Production* 11: 683–93.

UNDP (United Nations Development Programme). 2013. www.undp.org/content/undp/en/home/librarypage/environment-energy/inclusive-green-economy-approaches/

UNEP (United Nations Environment Programme). 2013. "Green Economy." www.unep.org/greeneconomy/AboutGEI/WhatisGEI/tabid/29784/Default.aspx

———. 2011. "Setting the Stage for a Green Economy Transition." www.unep.org/greeneconomy/Portals/88/.../GER_1_Introduction.pdf

———. 2009. "Global Green New Deal." Pittsburgh G20 Summit. www.unep.ch/etb/publications/Green%20Economy/G%2020%20policy%20brief%20FINAL.pdf

———. 1997. Papers from an Executive Seminar on the Role of Information Technology in Environmental Awareness-Raising, Policy-Making, Decision-Making, and Development Aid. 3 September.

Wals, Arjen. 2009. "Learning for a Sustainable World." Review of Contexts and Structures for Education for Sustainable Development. Wageningen University, The Netherlands: UNESCO Publications. www.unesco.org/education/justpublished_desd2009.pdf

WCED (World Commission on Environment and Development). 1987. *Our Common Future*. Oxford: Oxford University Press.

World Bank. 2013. "How Do You Measure Green Growth? World Bank and Partners Are Working on Indicators." www.worldbank.org/en/news/feature/2013/04/04/creating-global-green-growth-indicators

WRI (World Resources Institute). 2013. www.wri.org/stories/2011/04/qa-what-green-economy. 3 September.

CHAPTER 2

Cradle-to-Cradle and Innovation

Hussein ElKersh and Salah M. El-Haggar

The present chapter attempts to explore the Cradle-to-Cradle (C2C) concept, clarify it, and critically investigate its practicality for promoting sustainable product development. It also aims to suggest practical ways to apply the third principle of the C2C concept, respecting diversity. To investigate the practicality of the C2C concept, two points of criticism associated with its application are discussed and various challenges that can hinder its practical application are identified. Possible practical solutions to these challenges are also introduced.

The chapter discusses the practicality of achieving environmental sustainability through innovation. In order to link environmental sustainability and innovation, a good understanding of the essence of innovation is required. Thus, various definitions for innovation and eco-innovation as well as different types and approaches of innovation are explored. The factors that can affect eco-innovation and the challenges of integrating it into environmental sustainability are identified.

Cradle-to-Cradle Life Cycle
The biological and technical nutrients
In the C2C life cycle, materials flow in a closed loop without losing any resources through either a biological or a technical cycle/metabolism, as shown in table 2.1. Based on this concept, the word 'waste' is replaced by the word 'food.' This food can be directed either to the industry, to manufacture new equal or higher-quality products, in this case called technical nutrients, or to the living organisms, providing compost to be used as a soil conditioner or organic fertilizer, in this case called biological nutrients. Table 2.1 illustrates the differences between technical and

biological nutrients. It is important to note that the technical nutrients concept used in C2C considers products such as cars, computers, refrigerators, and washing machines as service products. This means that customers should purchase the service of these products for a defined use period, 'eco-leasing' instead of purchasing the products themselves. After this defined use period the products should be returned to the manufacturer to employ their complex materials as food for new products, and thus close the materials loop. The C2C concept has three guiding principles: waste equals food; use current solar income; and respect diversity. The idea of biological and technical nutrients helps to apply the first principle (i.e., waste equals food) (McDonough and Braungart, 2002).

Table 2.1. Biological and technical nutrients in the C2C concept (EPEA, 2012)

	Biological nutrients	Technical nutrients
Definition	Materials that biodegrade safely, provide food for living organisms, and make a closed loop.	Materials that cannot be returned to natural processes but can be completely recycled and reused in a closed loop.
After-use scenario	Used as compost, forming a nutrient basis for new natural resources.	Returned to the manufacturer, disassembled, and used to manufacture new products.
Example	Consumption products like certain packaging materials, clothing, and parts that undergo wear and tear such as car tires and brake discs.	Service products like washing machines, cars, and refrigerators.
Life cycle	The life cycle of these products is called the biological cycle.	The life cycle of these products is called the technical cycle.

Biological Nutrients   Technical Nutrients

Common misconceptions

Using the concept of closed cycles of resources, Orecchini (2007) introduced a new definition of sustainable development: "Sustainable development does not consume resources. It uses and reuses them, endlessly." However, to have a more accurate definition of sustainable development within this context, the word 'consume' should be replaced by the word 'waste.' Conserving natural resources through sustainable product development does not mean that resources are not consumed. Resources are consumed, but are then regenerated after their consumption.

The main goal of the C2C system is to promote sustainable product development through saving materials/resources by closing the materials loop. Once materials are used in a certain product they will circulate in a closed loop without losses. However, this does not mean that the C2C system eliminates the need for virgin materials (Braungart, 2012). Virgin materials are required for the manufacture of new products to meet the increasing needs associated with population growth. Virgin materials must be used since the demand for different resources already exceeds the waste generation rates (Bjørn, 2011; Bjørn and Hauschild, 2011). Thus, the C2C system does not eliminate the problem of resource depletion (Bjørn, 2011; Bjørn and Strandesen, 2011), but it mitigates this problem, conserving resources by using them in closed loops. This significantly reduces the amount of virgin materials required, defers the problem of resource depletion, and allows time to find alternative materials.

The claim that the C2C system targets short life cycle product design, so that nutrients can move quickly around in the closed loops to fuel the economy (Bjørn and Hauschild, 2011), is not completely correct. It is true that C2C allows for short life span products (Braungart, McDonough, and Bollinger, 2007). However, the defined use period for any product in the C2C system can be short or long, depending on the product itself. For example, it would not be appropriate to design a washing machine or a computer with a defined use period greater than ten years because this hinders innovations from reaching the market. Yet some products, such as windows, can have a defined use period of as long as twenty-five years. This defined use period gives manufacturers a timeframe for when they will get the materials back (Braungart, 2012).

Regarding the technical nutrients, the term 'recycling' is sometimes misunderstood. There are three sub-types of recycling: upcycling, recycling, and downcycling. These three sub-types correspond to products of higher, equal, or lower quality and/or value than the original ones. Downcycling simply slows the linear cradle-to-grave (C2G) life cycle rather than replacing it with the cyclic closed loop of C2C. Therefore, in the C2C system, only upcycling and recycling are accepted. However, a certain system can be considered as a sustainable system if there is a balance between downcycling and upcycling (Kane, 2002).

Eco-efficiency vs. eco-effectiveness

Eco-efficiency is a C2G-based approach that aims to minimize the volume, velocity, and toxicity of the material flow system, without altering its linear progression. Eco-effectiveness, on the other hand, is a C2C-based approach that aims to generate cyclical C2C metabolisms so materials can maintain their status as resources over time (Braungart, McDonough, and Bollinger, 2007). The C2C concept targets moving from eco-efficiency to eco-effectiveness, switching from open loops to closed loops. Table 2.2 illustrates the main differences between these two approaches.

One interesting issue regarding eco-efficiency and eco-effectiveness is the difference in the flow pattern of materials based on these two approaches. Eco-efficiency aims at reducing the large amounts of materials/resources used in the C2G open loop as shown in figure 2.1(a) and (b). However, it does not prevent such materials from ending up in landfills or incinerators. On the other hand, eco-effectiveness allows large amounts of materials to be used, as shown in figure 2.1(c), because such materials will be saved by flowing through the C2C closed loop (Braungart et al., 2008). For example, packaging materials can be used as much as desired to protect and differentiate products since these materials will then be conserved by circulating in a closed loop, becoming either biological or technical nutrients after their use (Newcorn, 2003). Since the C2C concept only defers the problem of resource depletion, a beneficial new approach may be introduced. The new approach combines the advantage of eco-efficiency, reducing material use, and the advantage of eco-effectiveness, conserving natural resources. This new approach can be called 'efficient eco-effectiveness,' or the 3 E's approach. In the 3 E's approach reduced amounts of materials are used, but within the C2C closed loops as shown in figure 2.1(d).

Table 2.2. Differences between eco-efficiency and eco-effectiveness
(Braungart, McDonough, and Bollinger, 2007; Wang and Côté, 2011)

	Eco-efficiency	Eco-effectiveness
Features	Doing more with less by reducing material, energy, and toxic substances and services; dematerialization	Regenerating with no dissipation by redesigning products and services for closing material flow systems; biodegradable or recyclable
Life cycle	C2G	C2C
Recycling type	Downcycling	Upcycling
Measures	Quantitative	Qualitative
Effect on ecological systems	Eco-efficiency aims to reduce harm to ecological systems. Eco-effectiveness	Eco-effectiveness strives to generate a beneficial effect for ecological systems. Eco-efficiency
Result	Doing better within a flawed system, resulting in a slower depletion of natural resources and fewer negative environmental impacts	Redesigning the entire industrial system, resulting in conservation of natural resources and creation of positive environmental impacts

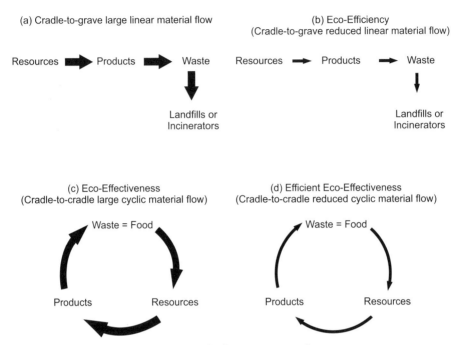

Figure 2.1. Material flow pattern comparisons

A hierarchy for the sustainability of material flow patterns is proposed in figure 2.2, the 3 E's approach at the top of the hierarchy representing the most sustainable material flow pattern.

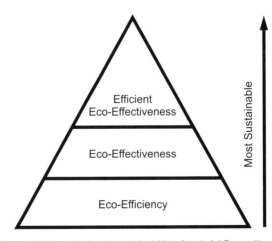

Figure 2.2. Hierarchy for the sustainability of material flow patterns

Respecting Diversity
Design flexibility

In the C2C system any product should be designed to be either a technical or a biological nutrient. Some products can be designed either way. Packages represent a good example. They can be designed as a technical nutrient if manufactured with recyclable materials such as polymers. Some polymers have already been tested and proved to be reusable up to ninety times with the same performance characteristics. Also, paper-based packages can be used as a technical nutrient if the inks used are designed such in a way that they can be washed out. This enables recyclers to produce white paper; at the same time the inks can be recovered and reused. Packages can also be designed as a biological nutrient by applying biodegradable materials or by adding more than 35% linear polyesters to PET (polyethylene terephthalate), which makes the whole material biodegradable (Newcorn, 2003).

Not only can products be designed as biological or technical nutrients, but also certain materials like polylactic acid (PLA)—which is used as packaging material for food and beverages—can be used either way. PLA is a bio-based polymer made out of lactic acid manometers, and it can be considered as a biological or a technical nutrient based on its end-of-life scenario. Chemical hydrolysis can be used to recycle PLA, making it a technical nutrient, while composting makes it a biological nutrient. Neither of the two metabolisms has higher priority than the other. This gives the designer flexibility to choose what he/she sees as more convenient (Bjørn and Hauschild, 2011). Nutrients that qualify to be used as biological or technical nutrients are called "twin nutrients" (Bjørn, 2011). Figure 2.3 illustrates a conceptual model of the technical and biological nutrients where the intersection represents twin nutrients.

Bjørn (2011) considers the presence of the twin nutrients in the C2C concept problematic because C2C does not give preference to the biological or the technical cycle. In certain cases, such as the use of PLA, the environmental benefits of the technical cycle might be far better than opting for the biological cycle. However, twin nutrients may be viewed as advantageous rather than problematic. They give the designer flexibility to choose whether to design the product as a biological or a technical nutrient based on the community where it will be used, thus respecting diversity.

One of the guiding principles in the C2C concept, respecting diversity, means that the unique features of each community or ecosystem

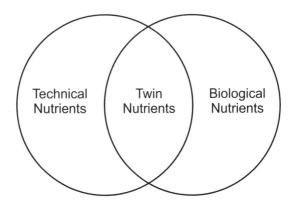

Figure 2.3. Twin nutrients (Bjørn, 2011)

should be taken into account when designing any product. By taking into consideration the differences among communities where products will be used, the flexibility of C2C product design may be practically applied. For example, in a community in a developed industrial country where recycling culture and technology exist, designing products as technological nutrients may be a practical approach to promote sustainable product development. Yet for a developing agricultural country, where this culture and technology do not exist, especially when considering upcycling, products designed as biological nutrients may represent the most practical approach. Including such products in the biological cycle is better than landfilling them and losing resources. This design flexibility helps apply the C2C concept to practically promote sustainable product development even if the environmental benefits of the technical cycle are more favorable than the biological cycle. However, such flexibility is not applicable to all products or materials, because they cannot all be designed to be used in both ways.

Delayed product differentiation

Many products are designed based on the 'one-size-fits-all' concept. This concept does not respect the diversity of ecosystems, resulting in universal product designs without consideration for the ecosystems where the products will be used. For example, detergents are almost the same the world over. To achieve a universal design based on the 'one-size-fits-all' concept, soap manufacturers use the worst-case scenario and manufacture detergents that have the same efficiency under the worst possible circumstances. The detergents used in places with soft water, like the northwestern region of the United States, are the same detergents used in places with hard water, like the southwestern region, despite the fact that soft water needs only small amounts of detergent to remove dirt. Manufacturers just add more chemicals in order to have a universal design. Such universal designs do not take into consideration the type of water available in each ecosystem (e.g., soft, hard, urban, or spring water), and also do not consider how the used water will be discharged, which can differ from one community to another (e.g., water can be discharged directly to fish-filled steams, or to sewage plants) (McDonough and Braungart, 2002).

Delayed product differentiation is a concept that has been used in supply chain management; however, it has never been considered within the context of environmental sustainability. The concept can be applied to manufacture products compatible with the ecosystems where they will be used. This could be a practical solution to the ecological problems associated with the 'one-size-fits-all' designs.

In the delayed product differentiation concept, the stage in the manufacturing process at which products are assigned individual different features is postponed. This helps to increase product variety with mass production. As an illustration, Hewlett-Packard implements this concept by conducting the localization steps for its computers and printers in its overseas distribution centers. These localization steps include installing the suitable power supplies for each country (Benjaafar, Heragu, and Irani, 2002). Figure 2.4 is a diagrammatic scheme that illustrates the delayed product differentiation concept. Each line of the products—P1, P2, and P3—represents a product that is compatible with

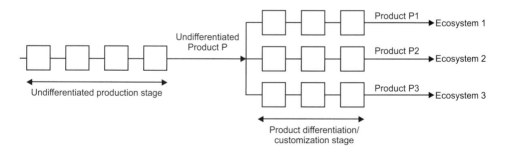

Figure 2.4. Diagrammatic scheme for delayed product differentiation

the ecosystem where it is going to be used. The differentiation/customization stage can be delayed even to the end stage, to be carried out in small manufacturing facilities in the countries or regions where the products will be used.

Companies usually offer a product portfolio that consists of families of similar products that have a small number of differentiating features. The concept of delayed product differentiation is applied to achieve this goal. For instance, different fragrances can be added to detergents using the concept of delayed product differentiation (Aviv and Federgruen, 2001). Applying this concept to manufacture detergents suitable to the type of water available in the ecosystems where these detergents will be used can help solve the ecological problems associated with the 'one-size-fits-all' detergents. Future research can identify to which products this concept, within its new context, can be effectively applied.

Criticism of C2C

Two main points of criticism have been identified regarding the concept of C2C: the mismanagement of biological nutrients and problems relating to energy considerations.

Mismanagement of biological nutrients

Braungart, McDonough, and Bollinger (2007) state that the quantity of emissions in the biological nutrients system is not a problem since they are emitted in the course of the production of healthy emissions, that is, "emissions containing biological nutrients which benefit the environment." On the other hand, Reijnders (2008) criticized this concept for two main reasons, the negative effect of the increase of biological nutrients and the fact that some biological nutrients can contain allelopathic substances detrimental to some organisms. Table 2.3 presents these two reasons along with illustrative examples.

One of the examples in table 2.3 states that adding biological nutrients to an ecosystem could be detrimental to some organisms while being healthy to others. This point can be supported by the study carried out by Norton et al. (2006). The study investigated the effects of different management inputs, fertilizer, seeds, and grazing patterns on plant biodiversity. It concluded that adding fertilizer and adventive seeds to infertile ecosystems can increase the vigor and abundance of some native species, but with an overall decline in the native species richness. This implies that the application of the biological nutrients

Table 2.3. Criticism of the biological nutrients concept, with examples
(Reijnders, 2008; Bjørn and Hauschild, 2011)

Criticism	Examples
Negative effects of the increase of biological nutrients	Increased emissions of CO_2, a biological nutrient released during oxidative biodegradation, have several negative impacts on both the environment and humans.
	Increased inputs in nitrogen and phosphorus cycles can have harmful effects. If excess nitrogen or phosphorus compounds (such as ammonia or phosphate) are directed to water bodies, they will result in eutrophication, which leads to excess production of algae, loss of oxygen, and loss of biodiversity in surface water ecosystems.
	Changes in the biogeochemical nitrogen cycle affect the atmosphere, leading to negative effects on the ozone layer and climate, which in turn affect human health and crop yields.
	The concentration of a certain biological nutrient that is optimal in one ecosystem could be toxic in others. Thus, adding biological nutrients to an ecosystem can be detrimental to some organisms while being healthy to some others, resulting in a violation of the third principle of C2C (respecting diversity).
Some biological nutrients contain allelopathic substances that can be detrimental to some organisms	Trees of the Prunus family, such as cherry trees, contain substantial amounts of allelopathic substances like phenols and cyanogenic glucosides, which may have harmful effects on some organisms if mulches made out of materials derived from these trees are applied.

concept in certain ecosystems may not lead to positive ecological outcomes because of the decline in the biodiversity values. Actually adding biological nutrients to ecosystems may not be beneficial except when these nutrients have already been degraded by human impacts in the first place (Bjørn and Strandesen, 2011).

In New Zealand a group of scientists who have a broad understanding of the biological processes associated with sustainable product
development were selected to assess the application of the C2C concept
from an ecological perspective. They pointed out that we need to be
careful when applying the biological nutrients concept as a simple solution to environmental sustainability problems (Reay, McCool, and
Withell, 2011). This study supports Reijnders' (2008) concerns
regarding the use of biological nutrients in the C2C life cycle.

On the other hand, Braungart (2012) states that most of these concerns are a result of the misunderstanding of the concept of the
biological nutrients used in the C2C system. Initially biological nutrients need to be carefully managed, which means that material balance
needs to be considered. Such balance defines where materials are actually used and how they are organized in the biological cycle. If certain
materials cannot be properly managed in the biological cycle they
should be used in the technical cycle. Furthermore, biological nutrients
should not be used without restriction or limitation because unlimited
uses may be inappropriate for rare materials such as phosphorus or
potassium. Use of these biological nutrients needs to be properly managed because they are finite. It is true that some activities in the
biological cycles might have some harmful consequences, but the question should be whether these consequences are reversible (for example,
no component of the cherry tree contains substances that accumulate in
the biological cycle, thus use of these components in the biological cycle
should be without adverse consequences).

Energy considerations

The C2C concept is criticized for not assessing the amounts of energy
needed during the life cycle of products (Bjørn and Hauschild, 2013;
Bjørn and Hauschild, 2011; Bjørn, 2011; Bjørn and Strandesen, 2011;
Bakker et al., 2010). It is true that C2C focuses on utilizing the current
solar income by using renewable energy resources during the manufacturing and assembly stage of products (MBDC, 2007), but the quantities
of energy used during the products' entire life cycle are not considered.
This approach is not reasonable because fossil fuels currently dominate
the energy sector and this is not expected to change in the near future
(Bjørn and Hauschild, 2011). For instance, utilizing solar income by
using photovoltaic cells instead of fossil fuels still faces many practical
challenges. Grid parity, when the price of solar electricity equals that of

the conventional sources for electricity on the power grid, is still a distant goal. This is especially true for some countries like China and India where electricity needs in the coming years are expected to increase significantly, and where electric power generation from coal is much cheaper (Lorenz, Pinner, and Seitz, 2008).

Since the C2C concept as a material-oriented tool does not evaluate the amount of required energy, there might be some situations where closing the material loop will be achieved at the expense of the energy required for transportation and recycling (Bjørn and Hauschild, 2011). This conundrum is quite clear in the criteria used in the C2C certification program founded in 2005 by McDonough Braungart Design Chemistry (MBDC). This certification program is a four-tiered approach consisting of Basic, Silver, Gold, and Platinum levels to reflect continuing improvement along the C2C trajectory, with Platinum representing the highest level (EPEA, 2012; MBDC, 2012). Despite the fact that energy is already one of the major criteria used in that certification system, the main focus of such a criterion is on using renewable energy sources only during the manufacturing and assembly stage. Neither the energy used by products, nor the energy required to disassemble and recycle products after their use, is assessed (MBDC, 2007). This may result in a product with a Platinum C2C certificate having negative environmental impacts. Bakker et al. (2010) evaluated the C2C concept based on several student design projects. They concluded that the C2C system barely addresses the energy consumption of products. Applying the C2C concept in realistic projects where renewable energy sources are not yet available can give misleading results. One of the projects, designing an electric kettle, showed that if the designer's focus is shifted from energy efficiency to material selection in order to close the resources loop, he/she may optimize the product from a C2C point of view, but actually worsen its performance over its entire life cycle.

On the other hand, Braungart (2012) argues that by selling the product service for a certain defined-use period instead of selling the product itself, the energy will be included in the cost of using this service. He suggests that the service provider take responsibility for energy use as part of the consumer contract. For example, when consumers buy the service of cars, they would obtain all fuel as part of the lease fee (Zlotin, Zusman, and Smith, 2002). Consumers would pay per kilometer and car manufacturers would be interested in fuel saving because this would make cars, as service products, cheaper, thus providing more profits.

This makes energy an integrated part in the C2C design. Yet an unanswered question in this regard is how the providers of certain service products, such as electric home appliances, can pay for the energy use of their products in a practical way.

Another argument for the C2C concept is that it looks at the environmental problems associated with excessive use of fuel (like the harmful emissions of CO_2, NO_x, SO_x) from a material rather than an energy point of view. For example, it suggests designing a vehicle that converts fuel to carbon black, water, and hydrogen. Carbon black can be used in many companies working in rubber and plastic manufacturing. These companies actually burn fuel just to produce carbon black. Furthermore, the hydrogen produced could be used to fuel the engine, while the water and carbon dioxide that are produced can be mixed to create carbonated water. Regarding NO_x, Braungart suggests that, instead of minimizing these oxides, the process should make as much of them as possible, collect them in an on-board "scrubber," and use them as fertilizer (Zlotin, Zusman, and Smith, 2002).

Eco-effective solutions aiming to make use of harmful car emissions may be promising, but they are still far from practical application. However, there already are some practical applications that have successfully created useful products from some types of harmful emissions. A good example is the industrial symbiosis of Kalundborg in Denmark, where sulfur emissions are used to manufacture liquid fertilizers and synthetic gypsum (El-Haggar, 2007). Another practical example is the innovative proprietary technology used by the California-based company Calera to manufacture useful products, including cement and concrete, out of CO_2 emissions (Calera, 2012).

Innovation and Eco-innovation

Integrating innovation into environmental sustainability can be a great help in saving the environment and promoting environmental sustainability. It is important to know the essence of innovation by exploring its definitions, approaches, types, and phases before starting to identify the factors that can affect its integration into environmental sustainability and the challenges that such integration may encounter.

Definitions

'Innovation' is a very popular word that has always been used by politicians, decision-makers, scientists, engineers, and many others. The word originally came from the Latin word *innovare*, that is, to make new, to renew, or to alter (Sarkar, 2007).

A review of literature shows that there are numerous definitions of innovation because this topic has been studied and researched by many different research communities with different audiences (Garcia and Calantone, 2002). There is some overlap between the various definitions, but their number and diversity mean that there is no authoritative definition. Many researchers claim that one of the challenges of innovation is the lack of a common interpretation, which undermines the clarity of the concept. Thus, it would be beneficial to have a generalized explanation adaptable to various disciplines (Baregheh, Rowley, and Sambrook, 2009).

Cumming (1998) presents various definitions of 'innovation' attributed to different authors since the 1960s and notes that the definition has changed slightly. In the 1960s and 1970s innovation was thought of as simply generating a new idea, as shown in the study carried out by Tinnesand (1973). In this study, gleaned from 188 publications, the word was interpreted as shown in table 2.4. More than half of the definitions during that period involved the creation or introduction of new ideas. Currently, most authors call the process of generating new ideas 'creativity.'

Table 2.4. Interpretation of the word 'innovation' (Tinnesand, 1973)

Interpretation	Percentage
The introduction of a new idea	36%
A new idea	16%
The introduction of an invention	14%
An idea different from existing ideas	14%
The introduction of an idea disrupting prevailing behavior	11%
An invention	9%

The definition of 'innovation' has recently been refined to include the concept of successful commercialization, which makes the word even more difficult to understand. This can be attributed to the increase in business competitiveness during the last thirty years (Cumming, 1998). For instance, the Organization for Economic Co-operation and Development (OECD) study on technological innovation defined innovation as "an iterative process initiated by perception of a new market and/or new service opportunity for a technology-based invention which leads to development, production, and marketing tasks striving for the commercial success of the invention" (OECD, 1991).

Baregheh, Rowley, and Sambrook (2009) tried to establish a multi-disciplinary definition of innovation by studying its definition in different disciplines, including economics, entrepreneurship, business, management, technology, science, and engineering. They were able to formulate a model that presents the 'essence' of innovation, irrespective of the organizational or disciplinary context, as shown in the diagrammatic definition in figure 2.5. The model consists of six components: stages, social, means, nature, type, and aim. The six components of the model describe the possible flow of the innovation process and indicate various starting points within that process. Any of these components can be the starting point of the flow of the innovation process, depending on the discipline. For example, focusing on the technical possibilities of a new product might be the starting point for engineers, whereas marketing specialists might concentrate on identifying potential new markets. Hence, they define innovation as "the multi-stage process whereby organizations transform ideas into new/improved products, service or processes, in order to advance, compete, and differentiate themselves successfully in their marketplace."

'Eco-innovation' is a relatively new term that currently has a wide use in the contexts of environmental sustainability. Thus, it is imperative to define it before discussing the relationship between environmental sustainability and innovation. Eco-innovation can be defined as the process of developing new ideas, behavior, products, and processes that contribute to a reduction in environmental burdens or to ecologically specified sustainability targets (Rennings, 2000). A more comprehensive definition of eco-innovation was given by Reid and Miedzinski (2008) as "the creation of novel and competitively priced goods, processes, systems, services, and procedures designed to satisfy human needs and provide a better quality of life for everyone with a whole-life-cycle minimal use of

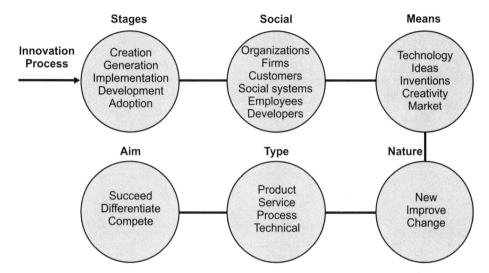

Figure 2.5. Diagrammatic definition of innovation (Baregheh, Rowley, and Sambrook, 2009)

natural resources (materials including energy and surface area) per unit output, and a minimal release of toxic substances." Such definitions can be considered within the context of eco-efficiency. Eco-innovations are sometimes referred to as environmental or sustainable innovations.

Innovation overview
Innovation approaches
There are two approaches to innovation: the closed and the open approach. The main difference between them lies in the way firms generate new ideas and bring them to the market.

The rule used in closed innovation is "Successful innovation requires control," which means that if a company wants to accomplish a successful innovation, it should depend entirely on itself. This was the dominant R&D philosophy during the twentieth century. Recently, however, firms have started to realize that the importance of the overall control used in closed innovation has started to diminish. Valuable ideas do not need to originate within the firm and the release of those ideas into the market is not necessarily accomplished by the firms' own activities (Herzog, 2011; Chesbrough, 2003). Table 2.5 illustrates the main differences between the two approaches; in the model of open innovation it is clear that the solid boundary between the firm and its environment becomes porous, allowing interchange between the firm and the environment.

Table 2.5. Closed vs. open innovation (adopted from Herzog, 2011; Chesbrough, 2003)

	Closed innovation	Open innovation
Concept	Firms must generate their own ideas, then develop, manufacture, market, distribute, and service the products themselves.	Firms use not only internal ideas and technologies as well as internal paths to market, but also external ideas, technologies, and paths to market, to advance their innovation projects.
Models		
Principles	A firm should hire the best and smartest people.	A firm does not need to employ all the smart people, but rather work with them inside and outside the firm.
	Profiting from innovative efforts requires a firm to discover, develop, and market everything itself.	Internal innovation activities are needed to claim some of the significant value which can be created by external innovation efforts.
	Being first to market requires that research discoveries originate within the firm. Being first to market also ensures that the firm will win the competition.	In order to win the competition, it is more important to have the better business model than to get to market first.
	Leading the industry in R&D investments results in coming up with the best and most ideas, and eventually in winning the competition.	Winning the competition does not require coming up with the best and most ideas, but to make the best use of internal and external ideas.
	Restrictive intellectual property (IP) management must prevent other firms from profiting from the firm's ideas and technologies.	Proactive IP management allows other firms to use the firm's IP. It also considers buying other firms' IP whenever it advances its own business model.

Encouraging open innovation leads to a more sustainable environment since it promotes spreading know-how, which is one of the major challenges that the transition toward environmental sustainability always encounters. Firms using that approach are able to cooperate to find innovative answers to tough environmental questions such as how to maximize renewable energy usage, how to minimize air pollution and have cleaner production technologies, and so on.

Sometimes new ideas and techniques are not viewed as promising in the firms that develop them, but might be very promising innovations for other firms. In the traditional closed innovation approach, these innovations are more likely to die; however, the new open innovation approach will make the best use of such innovations. A good example of this is Xerox and its Palo Alto Research Center (PARC). The researchers there developed many computer hardware and software technologies such as Ethernet and the Graphical User Interface (GUI). However, these were not viewed as promising innovations by Xerox, which was focusing on high-speed photocopiers and printers. Thus, these technologies were developed inside Xerox and commercialized by other companies. For example, Apple and Microsoft used the GUI in their Macintosh and Windows operating systems (Chesbrough, 2003).

Innovation types

Innovations fall into several categories, the four most common being architectural vs. component, radical vs. incremental, competence-enhancing vs. competence-destroying, and product vs. process innovation. Table 2.6 includes the descriptions of these categories as well as illustrative examples. In most cases, radical innovations are more likely to be competence-destroying, while on the other hand, most incremental innovations tend to be competence-enhancing (Darroch and McNaughton, 2002). It is worth mentioning, however, that most innovations take place in the incremental mode and eco-innovations are no exception (Hellström, 2007).

Innovation phases

There are four main phases that most innovative ideas should go through in order to be effective (Dormann and Holliday, 2002).
- The research phase, in which basic concepts are tested
- The development phase, in which the practical and economic aspects are checked for compatibility with the fundamental concepts

Table 2.6. Innovation types (adopted from Schilling, 2008)

Innovation category	Innovation type
Architectural vs. component innovation	Architectural
	Component
Radical vs. incremental innovation	Radical
	Incremental
Competence-enhancing vs. competence-destroying innovation	Competence-enhancing
	Competence-destroying
Product vs. process innovation	Product
	Process

Description	Example
Changes the overall design of a system or the way its components interact with each other	Transition from high-wheel bicycles to safety bicycles
Innovation to one or more components that does not significantly affect the overall configuration of the system	Changing ordinary bicycle seat material to gel-filled material for additional cushioning
Very new and different from prior solutions	Introducing new wireless telecommunication technology
Makes a relatively minor change from existing practices	Changing the configuration of a cell phone from one that has an exposed keyboard to one that has a flip cover
Builds on the firm's existing knowledge base	Developing Intel's microprocessors Pentium I, II, III and 4
Does not build on the firm's existing competences or renders them obsolete	Using calculators instead of slide rules
Embodied in the outputs of an organization (its goods or services)	Introducing the new hybrid electric vehicle
Innovations in the way an organization conducts its business (its techniques of production, marketing, or services) and oriented toward improving the efficiency of production	Reducing defect rates or increasing the quantity of a certain product that can be produced in a given time

- The demonstration phase, in which the best selected ideas are tested in pilot or full-scale form
- The commercialization phase, which includes manufacturing, licensing, and sales

It is worth noting, however, that innovations do not necessarily have to be complex and difficult to understand, but can be simple new techniques. In fact, when innovations are simple rather than complex, they are more likely to be adopted (Guerin, 2001). Some simple eco-innovations in companies may not necessarily pass through all four phases previously mentioned in order to be effective.

Factors affecting eco-innovation

Integrating innovation into environmental sustainability can be carried out by adopting and applying eco-innovation. However, sometimes the introduction of new technologies or eco-innovations is not welcomed by certain sectors of the society. Since eco-innovations are mostly driven by public policy rather than by free markets, they are likely to face resistance from established industries (Ekins, 2010). New technologies and innovations can also encounter great opposition and even cause social tension, for instance when introducing new machines that may need fewer workers or require existing workers to learn new skills (Vollenbroek, 2002). Such social aspects should be properly addressed when considering any innovation.

There are many other challenges that face innovators who try to promote the C2C concept through eco-innovative designs. One of the main challenges is determining a plan for sustainable living for a given group of people within the context of their environment. This will differ from one community to another, of course, depending on local customs, needs, and ecosystems. Thus, innovators must have a broad knowledge of science, art, engineering, communication, and human interaction; they must be aware of nature and of human needs, and seek to promote harmony between them (Stegall, 2006). Know-how is also an important factor in eco-innovations. Use of the open innovation approach can be of great importance in this regard.

The economic aspect is another major factor that greatly affects eco-innovation. Innovation for environmental sustainability requires additional costs in the form of new research facilities, long- and short-term research studies, and so on. While the ultimate aim is to produce environmentally friendly products and make use of waste to save natural resources, achieving profits is always a major concern. However, there is a common misconception that in order to be sustainable, huge costs are required with nothing in return except sustainability. This misconception is actually one of the main reasons why some industries and companies have not yet tried to take their first steps toward environmental sustainability. The profit motive, among other benefits, is one of the most important reasons for companies to support eco-innovations. These other benefits include saving money, increasing profits, reducing costs, and enhancing company reputation. Increased profits can come from innovating better products or creating new business (Nidumolu, Prahalad, and Rangaswami, 2009), while expenses can be saved by lowering disposal costs, reducing the material input costs per unit of output, and guaranteeing fewer environmental penalties and reduced liability insurance (Jackson and Clift, 1998). These cost savings will certainly cover any expenses required for innovation, but it will take some time. It will also create many job opportunities in, for example, R&D departments, and recycling and waste treatment facilities.

Despite the aforementioned benefits, some companies adopt environmental sustainability only because they face numerous external pressures such as the government imposing tough environmental regulations and policies. This is what happened in the 1980s when governments threatened large chemical firms with loss of their operating licenses if they did not adopt self-regulations through a set of environmental codes like pollution prevention (Hart, 2005). Companies have tried to get around this by relocating their industries to other parts of the world that do not have such policies. Local policies alone cannot result in protecting the environment, and this is where the importance of global agreements comes in (Vollenbroek, 2002). There are already many global agreements, but it should be noted that countries cannot be obligated to sign such agreements, and even after signing they have the right to withdraw (Barrett, 1994). While environmental regulations and policies can result in eco-innovation, Grubb and Ulph (2002) concluded that environmental policies may induce innovations that will lead to cleaner technologies, but that these environmental policies alone may not be sufficient for major eco-innovations. Technology as well as environmental policies are necessary for a significant environmental impact through environmental innovations.

Encouraging eco-innovation through shared organizational values, attitudes, and practices rather than focusing only on preventing, detecting, and punishing regulation violators can play a great role in promoting eco-innovation. A good example is the New Belgium Brewing Company (NBB), which became the first fully wind-powered brewery in the United States. The shared organizational values and attitudes prompted employees to vote to forfeit bonus pay to help the firm reduce its CO_2 emissions by 1,800 metric tons per year by investing in a new wind turbine system (Arnaud and Sekerka, 2010). It is very common to see a clear gap between the statements of companies and the employees' practice with regard to environmental sustainability and innovation. However, NBB is an example where there is no such gap. Encouraging creative problem solving, innovative products, and service ideas from the workforce can fill this gap. Proactive companies recognize the need to encourage these types of innovations because they know that innovation is crucial to the process of transformation into a sustainable enterprise. Still, even when they do have supportive environmental policies, line manager support is an important factor in promoting employee eco-innovation (Ramus, 2001).

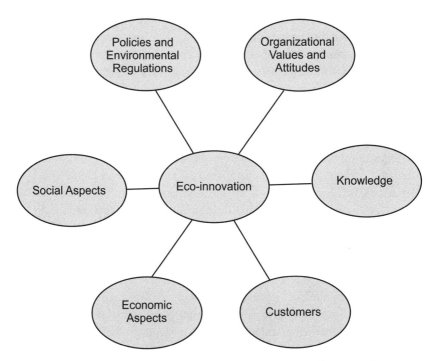

Figure 2.6. Factors affecting eco-innovation (ElKersh and El-Haggar, 2012)

Another important factor is the consumers themselves, who want to live in a pollution-free environment and save environmental resources for their children and grandchildren. Consumers have realized that their purchasing behavior can have a great impact on the environment. They check to see whether products are wrapped in recycled materials and select ecologically compatible products such as CFC-free hair spray. They are also willing to pay more for products that are environmentally friendly. In a study carried out in 1989, 67% of Americans were willing to pay 5% to 10% more for ecologically compatible products (Laroche, Bergeron, and Barbaro-Forleo, 2001). Figure 2.6 shows factors that encourage eco-innovation.

Challenges
Eco-innovation challenges

While there is an obvious agenda to integrate environmental sustainability into the innovation process, it is a long and difficult task. Moving toward a sustainable society will require both incremental and radical innovation in all aspects of society (Hines and Marin, 2004). There are three main challenges that may hinder the spread of environmental sustainability practices through innovation across different sectors.

The first challenge is the lack of awareness about the financial benefits that a company can achieve when turning to environmental sustainability. This is more obvious in the developing countries. It is the role of academic institutions, governments, NGOs, and the media to make their societies more aware of the great financial benefits of applying environmental sustainability practices.

The second challenge is the time it takes to begin realizing those financial benefits. This time frame differs from one sector to another depending on the activity and size of each sector. It may take a considerable time, especially if the innovation process will need to go through all four of the innovation phases previously discussed. What makes this more challenging is the fact that only one out of several thousand ideas results in a successful new product. Many innovation projects do not result in technically feasible products, and even for those that do, many fail to earn a commercial return (Schilling, 2008). Across most industries, it appears to require 3,000 raw ideas to produce one substantially new, commercially successful industrial product (Stevens and Burley, 1997). This is illustrated by the innovation funnel shown in figure 2.7.

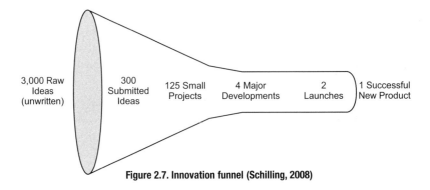

3,000 Raw Ideas (unwritten)   300 Submitted Ideas   125 Small Projects   4 Major Developments   2 Launches   1 Successful New Product

Figure 2.7. Innovation funnel (Schilling, 2008)

Innovation becomes even more challenging during economic recession, since it is very difficult for a company to spend money on something that will need some time to yield profit at a time when it needs to make cost reductions almost everywhere. It is interesting, however, that several cost-reduction actions are in the first place sustainable actions. These include reducing the number of company cars used for transporting employees and carpooling or even using public transportation instead, which reduces harmful carbon emissions.

The third challenge is the know-how. Sometimes firms are eager to achieve environmental sustainability and solve the environmental problems associated with their products and industries but don't know how to do so technically. However, an open innovative approach by developing collaborations between R&D departments in different firms can be a key solution.

C2C challenges

There are various challenges that can hinder the practical application of the C2C concept in promoting sustainable product development. These challenges are related to four main aspects: life cycle assessment (LCA), upcycling, eco-leasing, and materials. Table 2.7 presents a summary of the identified challenges and the practical solutions that have been suggested. These challenges and solutions will be discussed in detail.

Life cycle assessment

Life cycle assessment (LCA) is a methodology for evaluating the environmental impacts related to any product. LCA assesses these impacts for the entire life cycle of products starting from collecting raw materials,

Table 2.7. Identified challenges and suggested practical solutions

	Challenges	Suggested solutions
LCA	Unsustainable product development (when depending on LCA or C2C alone)	Use LCA and C2C as complementary tools to assess the environmental impacts of products during the design phase
	C2G LCA-ISO 1404x standards	Change LCA-ISO 1404x standards from C2G to C2C
Upcycling	Directing materials to landfills and incinerators through continuous downcycling	Use of the cascade principle
	Knowledge of the exact materials used in each component	Apply the concept of intelligent materials pooling and banking
		Involve a third party
	Technology and innovation	Adopt the open innovation approach
	Difficulty of disassembling and recycling some products	Design for recycling and disassembly (for example, simplified modular design)
Eco-leasing	Lack of reversed logistics for returning products to their manufacturers Orphaned and imported products	Adopt the pooled take-back approach
	Not practical for some products	Design such products as biological nutrients
	Transition to Product Service System (PSS)	Address the uncertainties of the PSS system (for example, the readiness of manufacturers to adopt this system and the readiness of customers to accept it)
Materials	Toxic/hazardous materials	Make sure that toxic materials are circulated in completely closed technical cycle (until other alternatives become available)
	Current tax system	Spread taxes more evenly between labor and raw materials
	Biological–technical and technical–technical nutrients mix	Develop biological–biological nutrients mix

until all residuals are returned to the earth for 'disposal' based on the C2G life cycle (El-Haggar, 2007). The first documented LCA was carried out by Coca-Cola in the 1960s to compare the impacts of cans and glass bottles (Knight, 2009). In contrast with LCA, C2C can be used as a tool not only to protect the environment, but also to save natural resources by closing the materials loop and switching from eco-efficiency to eco-effectiveness. Table 2.8 shows the main differences between C2C and LCA.

Bjørn (2011) evaluated the environmental sustainability of a C2C-certified mineral-based paper product, TerraSkin, by using LCA. The results were then compared with an eco-efficient paper alternative, cellulose-based paper with comparable properties. The results showed that

Table 2.8. Comparison between C2C and LCA (Hupperts et al., 2011)

	C2C	LCA
Philosophy	It is possible to design products for a positive impact on people, the environment, and economic profit	All products pollute: they all require extraction of raw materials and there is always some form of waste left over
Approach	Eco-effectiveness	Eco-efficiency
Design aids	Use 3 guiding principles to establish a clear direction (waste equals food, use the current solar income, and respect diversity)	Use hot spots (life cycle elements with the biggest negative environmental impact) to set priorities for improvements
Environmental impact	Maximizing positive effects on people, their environment, and conservation of resources	Minimizing the negative environmental impacts by trying to reduce pollution and material use
Footprint	Develop a positive beneficial footprint	Measure the footprint and let designers decide how to handle it

although TerraSkin has significantly lower impacts in toxicity-related categories, as well as land occupation and fresh water consumption, it has a significantly higher primary energy demand throughout its life cycle. This demonstrates that a C2C product is not guaranteed to be more sustainable than eco-efficient alternatives when applying LCA as a measuring tool (Bjørn, 2011; Bjørn and Hauschild, 2013). It is recommended that LCA and C2C be used as complementary approaches in product design. For example, LCA can be used to map significant energy flows in the life cycle of products, while C2C can be used to close the open resources loop (Bakker et al., 2010). Prendeville, O'Connor, and Palmer (2011) used C2C and LCA as eco-analysis tools to comprehensively evaluate the environmental performance of a product in small to medium enterprises (SMEs). They also concluded that C2C and LCA can be used in a complementary manner for sustainable product development.

Another issue related to C2C and LCA is that C2C is about quality, while LCA, as a measuring tool, is about quantity. The qualitative analysis used in C2C provides a very flexible framework, which has its own weaknesses and strengths. The strength is that it persistently seeks to achieve positive impacts through eco-effectiveness, which strongly encourages innovation. However, such a flexible framework can be considered as a point of weakness because of the lack of well-established methodological procedures (Bakker et al., 2010).

The International Organization for Standardization (ISO) has developed a series of international standards to cover LCA. The first edition of ISO 14044 *(Life Cycle Assessment: Requirements and Guidelines)*, released in 2006, together with the second edition *(Life Cycle Assessment: Principles and Framework)*, released in the same year, are currently used to cover LCA instead of the four standards: ISO 14040:1997, ISO 14041:1998, ISO 14042:2000, and ISO 14043:2000 (ISO, 2006a; 2006b). However, all LCA-ISO 1404x standards are based on the C2G life cycle, which does not conserve natural resources. The standards actually encourage the safe disposal of products that result in resource waste. Thus, complying with regulations based on these standards protects the environment but without conservation of resources. ElKersh and El-Haggar (2012) introduced the concept of Beyond Compliance (BC), by which an organization would not only seek compliance with regulations, but would also strive for sustainable development. With that concept, compliance with regulations can be thought of as only the basic initiative for the journey toward environmental sustainability. It would be very beneficial to change the LCA-ISO 1404x standards from C2G to C2C, not only to protect the environment, but also to save natural resources.

Upcycling

As previously explained, only upcycling and recycling are acceptable in the C2C concept. Downcycling is usually associated with losing some valuable materials, resulting in the production of lower-quality or lower-value products. Moreover, continuous downcycling of products does not save resources and materials; it only defers the time when products will eventually end up in landfills or incinerators.

Upcycling can be a major challenge in terms of technology as well as know-how. It is quite difficult to upcycle any product without knowing the exact materials used in each of its components. Therefore, the involvement of the manufacturers as well as the suppliers of the materials is critical. Upcycling requires that innovation be integrated with sustainable product development. Several innovative products can be manufactured by upcycling the unrecyclable wastes, such as plastic rejects from municipal solid waste. Such products include bricks, interlocks, manhole covers, table toppings, and road ramps (El-Haggar, 2007). Another example of an innovative product manufactured by the upcycling of what used to be called waste is the concept phone "Remade," manufactured by the mobile phone maker Nokia. This phone is made almost entirely of recycled materials such as aluminum cans, plastic bottles, and even old car tires (McFedries, 2008).

As previously discussed, the open innovation approach, in which companies use external ideas, technologies, and external paths to accomplish their innovation projects through collaboration, helps to meet the challenge of know-how and thus promotes environmental sustainability. Similarly, the concept of intelligent materials pooling and banking introduced by McDonough and Braungart (2003) may help meet the upcycling challenge through the sharing of information among participating companies (Braungart, McDonough, and Bollinger, 2007; McDonough and Braungart, 2003). In certain cases, the involvement of a third party external to the material flow system can represent another solution. This is especially true if the company that supplies materials to the manufacturing companies wants, for competitive reasons, to keep the specifics of their formulations confidential. In such a case, this third party can have the role of ensuring that all actors within the material flow metabolism have access to the information that they need, while also ensuring that the information remains proprietary (Braungart, McDonough, and Bollinger, 2007).

Designing for recycling and disassembly can be a great help in meeting the challenge of upcycling, making it easier to recycle, upcycle, and/or disassemble any product after its usage. Disassembly can be expedited for some products through simplified modular designs. Standard fasteners and easy-to-release disassembly methods can be used in these designs to enable the manufacturers to easily upgrade or add options to the product (upcycling) after its defined use period (Zlotin, Zusman, and Smith, 2002). The "Mirra Chair" designed by Herman Miller is a good example of applying the C2C concept in a practical way by increasing the recyclability of products and facilitating the disassembly process (Rossi et al., 2006).

The challenges associated with upcycling are especially difficult in developing countries. Even developed countries find it difficult to upcycle products. For instance, in 2003 the Netherlands generated 24,967 million tons of waste from construction and demolition; of this, 23,977 million tons were downcycled, with 90% used as road base (Durmisevic, 2008). Timber, which is used in the construction of windows, usually ends up in lower-value products such as chipboard or animal bedding (Asif, Muneer, and Kubie, 2005). The cascade principle, in which materials are designed to suit both the biological and the technical cycles, may furnish part of the solution to this challenge. Bio-based material may go through "a cascade of consecutive uses of lesser and lesser properties" beginning in the technical cycle, but ending in the biological cycle (Bjørn and Hauschild, 2013). Office paper is a good example. It can be repeatedly recycled (up to seven times) while keeping its long fibers, which enables the recycler to use it repeatedly for the same purpose. The next stage—when the fibers become too short to be used as office paper—includes different reuse scenarios such as cardboard, tissues, or toilet paper. Finally, the material is returned to the ecosystem as a biological nutrient through ash, compost, or waste water (NL Agency, 2011).

Downcycling can be used as a temporary, short-term, eco-efficient solution when no other options except landfilling or incineration (without cascading and recovery of beneficial ash) are available. This will save the materials and resources required to manufacture the downcycled products until an eco-effective solution becomes available. For example, manufacturing insulation materials from downcycled high-quality plastic products results in saving the resources required for manufacturing these insulation materials, so that virgin materials can be used to manufacture new high-quality plastic products. As previously explained, using virgin materials is not a violation of the C2C concept.

Eco-leasing

To implement the technical nutrients concept, service products such as cars, computers, refrigerators, and washing machines should be returned to their manufacturers after a certain defined use period for disassembly to be upcycled or recycled. These service products would be eco-leased instead of sold. In practical application, this aspect of the technical nutrients concept is a major challenge for the C2C process. It is not easy to establish a reverse logistics system to return products to their manufacturers. The availability of take-back networks represents a barrier (Jacques, Agogino, and Guimarães, 2010). At the same time, recycling such products without the participation of their original manufacturers might result in downcycling due to lack of information about the exact materials used in each component in the products. In several cases C2C managed to recover materials, yet full integration with the existing forward supply chain systems is still challenging (Lie, 2010). Consumer cooperation may be especially challenging with some products. For instance, returning apparel items for recycling is not a common practice, and take-back initiatives in this area have met only limited success (Jacques, Agogino, and Guimarães, 2010). Designing such products as biological nutrients may be a more practical approach.

Table 2.9 summarizes some typical manufacturer collection methods for retrieving used products, along with illustrative examples. Some methods involve the retailer or a third party in collecting products and returning them to the manufacturer.

The challenge becomes even more complicated in the case of imported products, since it is not economically feasible to collect and return them to their manufacturers. According to some Extended Producer Responsibility (EPR) schemes, the importer has the full responsibility; however, the importers may not be properly equipped to demanufacture these products, especially if the upcycling option is considered. Orphaned products, those whose manufacturer no longer exists, represent another challenge in this regard. One solution for the reversed logistics challenge, as well as the challenge imposed by

Table 2.9. Technical nutrient collection methods (Kumar and Putnam, 2008)

Collection method	Diagrammatic presentation	Examples of products
Manufacturer collects from consumer; retailer is not involved	Manufacturer → Retailer → Consumer	Xerox and Canon use prepaid mailboxes Hewlett Packard picks up from local offices
Retailer collects from consumer and manufacturer buys back from retailer	Manufacturer → Retailer → Consumer	Kodak single-use cameras are returned to a retailer for developing and Kodak "buys back" Refrigerators and televisions are traded in
Third party collects used products from the consumer	Manufacturer → Retailer → Consumer, Third Party	Old car dealerships (Ford, GM, Chrysler, BMW, Fiat, and Renault), junkyards, recycling centers, and disassemblers sell recovered parts or materials back to manufacturer

imported and orphaned products, is the pooled take-back approach, in which the responsibility for the products is assigned to consortia of manufacturers who produce the same category of products. These consortia represent a Producer Responsibility Organization (PRO) that has its own demanufacturing facilities. Such an approach makes it easier to develop simpler collecting systems and solve the issue of imported and orphaned products (Spicer and Johnson, 2004). However, the manufacturers would have to share all the information regarding the exact materials used in product components with the PRO in order to avoid downcycling.

The notion of eco-leasing implies that the technical nutrient concept will require transition to a Product Service System (PSS) which focuses on the 'sale of use' instead of the 'sale of product.' However, in order to have a feasible PSS many uncertainties have to be addressed. Such uncertainties include the readiness of manufacturers to adopt this system and the readiness of customers to accept it. One of the main barriers to adopting PSS is the significant change in the system of profit-making for manufacturers. With limited experience in pricing and offering such products, manufacturers may find transition toward PSS very challenging (Baines et al., 2007; Mont, 2002).

Materials

Materials present another hurdle for the practical application of the C2C concept. There are four main challenges in this regard: the biological–technical nutrients mix, the technical–technical nutrients mix, toxic and hazardous materials, and the current tax system.

As previously discussed, products should be designed to be either biological or technical nutrients; if a product combines both technical and biological nutrients, they should be clearly marked and easily separable (MBDC, 2007). This represents a challenge for product designers, especially when working on products that contain mixes of both nutrients. This notion also implies that mixing of technical nutrients should be very limited so that one product does not contain too many different types of technical nutrients, thus complicating recycling (Bjørn and Hauschild, 2011). A good example is car recycling. Currently there is no technology to separate the polymer and paint coatings from car metals before recycling. Even if a car was designed to be completely disassembled, it would not be technically feasible to close the loop to retrieve its high-quality steel (McDonough and Braungart, 2002). The recycling of the fiber-reinforced composites that are used in many products also presents many challenges. Chemical recycling and regrinding methods are used in this regard, but the regrinding method is not recommended from a C2C point of view. This is because most of the retrieved materials have a lower quality than virgin materials, which results in downcycling the retrieved scraps for much less demanding applications (Song, Youn, and Gutowski, 2009). This is not acceptable from a C2C point of view; however, these technical–technical mixes can have lower environmental impacts than a product designed based on a purely technical C2C nutrient. For instance, using composite materials in car manufacturing instead of homogenized materials leads to a substantial reduction in the weight of cars, resulting in the use of less fuel, hence reducing harmful emissions (Bjørn, 2011; Bjørn and Hauschild, 2011). This deficiency in the C2C concept is attributed to the fact that it does not account for the energy use of products during their life cycle, as previously discussed.

The challenge of the technical–technical nutrients mix can be solved by replacing it with a suitable biological–biological nutrients mix. Such a biological mix of nutrients does not represent a problem at the end of their life cycle because they are completely biodegradable. The production of bio-composites that are competitive with synthetic composites is a developing area of research; there are ongoing investigations into the combination of bio-fibers such as kenaf, hemp, flax, jute, henequen, pineapple leaf, and sisal with polymer matrices from renewable resource–based biopolymers such as cellulosic plastics, polylactides, starch plastics, polyhydroxyalkanoates (bacterial polyesters), and soy-based products (Mohanty, Misra, and Drzal, 2002). Figure 2.8 shows a closed biological cycle using biopolymers as biological nutrients.

Although bio-based polymers can solve the problem of the open-resource loop, incorporating them into the C2C concept may present many challenges, including the current unsustainable practices in producing these polymers. The bio-based feedstocks used to manufacture

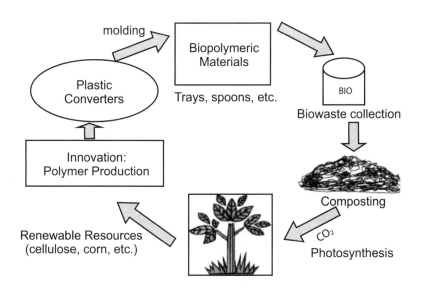

Figure 2.8. Closed biological cycle using biopolymers as biological nutrients
(Mohanty, Misra, and Drzal, 2002)

bio-based polymers are generally grown using industrial agricultural production methods that utilize significant amounts of toxic pesticides, polluting both water and soil and negatively impacting wildlife habitat. Significant amounts of water, energy, hazardous chemicals/additives, and genetically modified organisms are used to process these polymers. Infrastructure for composting bio-plastics is not widely available yet either (Álvarez-Chávez et al., 2012). It is even difficult to assess the environmental impacts through LCA for biodegradable plastics like PLA, especially if composting is considered as the end-of-life scenario. Data required for this assessment are lacking (Madival et al., 2009).

Toxic and hazardous materials pose another hurdle. C2C aims to completely avoid using such materials; still, some toxic substances have proven to be essential in the manufacturing of certain products, such as insulated windows. In C2C one prerequisite for the use of such substances is to make sure that the technical cycles in which these substances circulate are completely closed (EPEA, 2012). However, some toxic and hazardous materials cannot fit in the biological or technical metabolisms. These materials are called 'unmarketables.' Until solutions for their detoxification can be found, the unmarketables should be stored in safe repositories (McDonough and Braungart, 2002).

Other practical challenges to the application of the C2C concept with regard to materials are current tax systems, which put a heavy burden on labor while the use of raw materials is relatively cheap. From a C2C perspective, this does not encourage closing the open resources loop. Spreading taxes more evenly between labor and raw materials would lead to a relative rise in the price of raw materials, generating substantial financial incentive to close gaps in the cycle of materials. This would certainly encourage the application of the C2C concept (Hupperts et al., 2011).

Questions
1. What does the cradle-to-cradle life cycle mean?
   Why it is recommended to switch to this life cycle?
2. Compare the following:
   Cradle-to-cradle and cradle-to-grave life cycles
   Biological and technical nutrients
   Eco-effectiveness and eco-efficiency
   Open and closed innovation

3. Briefly explain the main points of criticism of the cradle-to-cradle concept.
4. Explain the differences between C2C as a tool and LCA. What are the advantages of using each of them?
5. What are the factors that can affect adopting and applying eco-innovation?
6. What are the challenges that are usually encountered in integrating innovation into environmental sustainability? Suggest solutions to meet these challenges.
7. Discuss the challenges that can hinder the practical application of the cradle-to-cradle concept and the possible solutions for such challenges. Which of these challenges do you think is the most difficult? Why?

References

Álvarez-Chávez, C.R., S. Edwards, R. Moure-Eraso, and K. Geiser. 2012. "Sustainability of Bio-based Plastics: General Comparative Analysis and Recommendations for Improvement." *Journal of Cleaner Production* 23, no. 1: 47–56.

Arnaud, A., and L.E. Sekerka. 2010. "Positively Ethical: The Establishment of Innovation in Support of Sustainability." *International Journal of Sustainable Strategic Management* 2, no. 2: 121–37.

Asif, M., T. Muneer, and J. Kubie. 2005. "Sustainability Analysis of Window Frame." *Building Services Engineering Research and Technology* 26, no. 1: 71–87.

Aviv, Y., and A. Federgruen. 2001. "Design for Postponement: A Comprehensive Characterization of Its Benefits under Unknown Demand Distributions." *Operations Research* 49, no. 4: 578–98.

Baines, T., H. Lightfoot, S. Evans, A. Neely, R. Greenough, J. Peppard, R. Roy, E. Shehab, A. Braganza, A. Tiwari, J. Alcock, J. Angus, M. Bastl, A. Cousens, P. Irving, M. Johnson, J. Kingston, H. Lockett, V. Martinez, P. Michele, D. Tranfield, L. Walton, and H. Wilson. 2007. "State-of-the-art in Product-service Systems: Proceedings of the Institution of Mechanical Engineers, Part B." *Journal of Engineering Manufacture* 221, no. 10: 1543–52.

Bakker, C.A., R. Wever, Ch. Teoh, and S. De Clercq. 2010. "Designing Cradle-to-Cradle Products: A Reality Check." *International Journal of Sustainable Engineering* 3, no. 1: 2–8.

Baregheh, A., J. Rowley, and S. Sambrook. 2009. "Towards a Multidisciplinary Definition of Innovation." *Management Decision* 47, no. 8: 1323–39.

Barrett, S. 1994. "Self-enforcing International Environmental Agreements." *Oxford Economic Papers*, 878–94. Oxford: Oxford University Press.

Benjaafar, S., S.S. Heragu, and S.A. Irani. 2002. "Next Generation Factory Layouts: Research Challenges and Recent Progress." *Interfaces* 32, no. 6: 58–76.

Bjørn, A. 2011. "Cradle to Cradle and Environmental Sustainability: A Critical Appraisal." Master's thesis, Department of Management Engineering, The Technical University of Denmark.

Bjørn, A., and M.Z. Hauschild. 2013. "Absolute versus Relative Environmental Sustainability." *Journal of Industrial Ecology* 17, no. 2: 321–32.

———. 2011. "Cradle to Cradle and LCA: Is There a Conflict?" In *Glocalized Solutions for Sustainability in Manufacturing: Proceedings of the Eighteenth CIRP International Conference on Life Cycle Engineering, 2–4 May 2011, Technische Universität Braunschweig, Germany*, edited by J. Hesselbach and C. Herrmann. Berlin and Heidelberg: Springer-Verlag.

Bjørn, A., and M. Strandesen. 2011. "The Cradle to Cradle Concept: Is It Always Sustainable?" Poster session presented at the Life Cycle Management (LCM) Conference: Towards Life Cycle Sustainability Management, 28–31 August 2011. Berlin.

Braungart, M. 2012. Personal communication.

Braungart, M., P. Bondesen, A. Kälin, and B. Gabler. 2008. "Specific Public Goods for Economic Development: With a Focus on Environment." Public Goods for Economic Development Compendium of Background Papers. Vienna: United Nations Industrial Development Organization.

Braungart, M., W. McDonough, and A. Bollinger. 2007. "Cradle-to-Cradle Design: Creating Healthy Emissions: A Strategy for Eco-effective Product and System Design." *Journal of Cleaner Production* 15, no. 13–14: 1337–48.

Calera. 2012. www.calera.com.

Chesbrough, H.W. 2003. "The Era of Open Innovation." *MIT Sloan Management Review* 44, no. 3: 35–41.

Cumming, B.S. 1998. "Innovation Overview and Future Challenges." *European Journal of Innovation Management* 1, no. 1: 21–29.

Darroch, J., and R. McNaughton. 2002. "Examining the Link between Knowledge Management Practices and Types of Innovation." *Journal of Intellectual Capital* 3, no. 3: 210–22.

Dormann, J., and C. Holliday. 2002. "Innovation, Technology, Sustainability and Society." World Business Council for Sustainable Development. www.bvsde.paho.org/bvsacd/cd30/society.pdf

Durmisevic, E. 2008. "The Netherlands, Construction Materials Stewardship: The Status Quo in Selected Countries." CIB report 318, edited by J.B. Storey. www.iip.kit.edu/downloads/CIB_Publication_318.pdf

Ekins, P. 2010. "Eco-innovation for Environmental Sustainability: Concepts, Progress and Policies." *International Economics and Economic Policy* 7, no. 2–3: 267–90.

ElKersh, H., and S. El-Haggar. 2012. "Towards Sustainability through Innovation." *Interdisciplinary Environmental Review* 13, no. 2/3: 187–201.

EPEA (Environmental Protection Encouragement Agency). 2012. www.epea-hamburg.org

Garcia, R., and R. Calantone. 2002. "A Critical Look at Technological Innovation Typology and Innovativeness Terminology: A Literature Review." *Journal of Product Innovation Management* 19, no. 2: 110–32.

Grubb, M., and D. Ulph. 2002. "Energy, the Environment, and Innovation." *Oxford Review of Economic Policy* 18, no. 1: 92–106.

Guerin, T.F. 2001. "Why Sustainable Innovations Are Not Always Adopted." *Resources, Conservation and Recycling* 34, no. 1: 1–18.

El-Haggar, S. 2007. *Sustainable Industrial Design and Waste Management: Cradle to Cradle for Sustainable Development.* Boston: Elsevier Academic Press.

Hart, S.L. 2005. "Innovation, Creative Destruction and Sustainability." *Research Technology Management* 45, no. 5: 21–27.

Hellström, T. 2007. "Dimensions of Environmentally Sustainable Innovation: The Structure of Eco-innovation Concepts." *Sustainable Development* 15, no. 3: 148–59.

Herzog, P. 2011. *Open and Closed Innovation: Different Cultures for Different Strategies.* 2nd rev. ed. Wiesbaden: Gabler Verlag, Springer Fachmedien Wiesbaden.

Hines, F., and O. Marin. 2004. "Building Innovations for Sustainability: Eleventh International Conference of the Greening of Industry Network." *Business Strategy and the Environment* 13, no. 4: 201–208.

Hupperts, P., F. Embrechts, J. Crol, A. Bor, and I. Nijs. 2011. "Cradle to Cradle Pays Off! Companies of the C2C Community about Their Experiences and Lessons Learned." The Netherlands: Knoops. www.theterrace.nl/assets/cms/File/Booklet_Learning_Community_C2C-Cradle_to_Cradle_pays_off!.pdf

ISO. 2006a. *ISO 14044: 2006 Environmental Management, Life Cycle Assessment, Requirements and Guidelines.* Geneva: ISO International Standards Organization.

————. 2006b. *ISO 14040: 2006 Environmental Management: Life Cycle Assessment, Principles and Framework*. Geneva: ISO International Standards Organization.

Jackson, T., and R. Clift. 1998. "Where's the Profit in Industrial Ecology?" *Journal of Industrial Ecology* 2, no. 1: 3–5.

Jacques, J., A. Agogino, and L. Guimarães. 2010. *Sustainable Product Development Initiatives in the Footwear Industry Based on the Cradle to Cradle Concept: Proceedings of the ASME 2010 International Design Engineering Technical Conferences and Computers and Information, in the Fifteenth Design for Manufacturing and the Lifecycle Conference (DFMLC) IDETC/CIE 2010, 15–18 August 2010, Montreal, Quebec, Canada*. American Society of Mechanical Engineers.

Kane, G. 2002. "Going Green for a Living: A Practical Guide (Sustainable Industrial Development)." *Engineering Management Journal* 12, no. 5: 214–19.

Knight, A. 2009. *Hidden Histories: The Story of Sustainable Design*. ProQuest Discovery guides, June. csadb102.csa.com/discoveryguides/design/review.pdf

Kumar, S., and V. Putnam. 2008. "Cradle to Cradle: Reverse Logistics Strategies and Opportunities across Three Industry Sectors." *International Journal of Production Economics* 115, no. 2: 305–15.

Laroche, M., J. Bergeron, and G. Barbaro-Forleo. 2001. "Targeting Consumers Who Are Willing to Pay More for Environmentally Friendly Products." *Journal of Consumers Marketing* 18, no. 6: 503–20.

Lie, K. 2010. "Cradle to Cradle: Incorporating Closed-loop Material Chains in the Industry." Master's thesis, Economics and Informatics, Erasmus University Rotterdam.

Lorenz, P., D. Pinner, and T. Seitz. 2008. "The Economics of Solar Power." *The McKinsey Quarterly*. ceodifference.org/clientservice/sustainability/pdf/Economics_of_Solar.pdf

Madival, S., R. Auras, S.P. Singh, and R. Narayan. 2009. "Assessment of the Environmental Profile of PLA, PET and PS Clamshell Containers Using LCA Methodology." *Journal of Cleaner Production* 17, no. 13: 1183–94.

MBDC. 2012. "McDonough Braungart Design Chemistry." www.mbdc.com/
————. 2007. "Cradle to Cradle Certification Program." www.greenbiz.com/sites/default/files/document/Outline_CertificationV2.1_Final.pdf

McDonough, W., and M. Braungart. 2003. "Intelligent Materials Pooling: Evolving a Profitable Technical Metabolism through a Supportive Business Community." *Green @ Work Magazine*, March/April. www.greenatworkmag.com/gwsubaccess/03marapr/perspect.html

————. 2002. *Cradle to Cradle: Remaking the Way We Make Things*. New York: North Point Press.

McFedries, P. 2008. "E-cycling E-waste: Recycling Old Words to Reprocess Old Electronics." *IEEE Spectrum*. spectrum.ieee.org/green-tech/conservation/ecycling-ewaste

Mohanty, A.K., M. Misra, and L.T. Drzal. 2002. "Sustainable Bio-composites from Renewable Resources: Opportunities and Challenges in the Green Materials World." *Journal of Polymers and the Environment* 10, no. 1/2: 19–26.

Mont, O.K. 2002. "Clarifying the Concept of Product–Service System." *Journal of Cleaner Production* 10, no. 3: 237–45.

Newcorn, D. 2003. "Cradle-to-Cradle: The Next Packaging Paradigm?" *Packaging World*, May: 62–68. www.globalcommunity.org/business/c2c1.pdf

Nidumolu, R., C.K. Prahalad, and M.R. Rangaswami. 2009. "Why Sustainability Is Now the Key Driver of Innovation." *Harvard Business Review* 87, no. 9: 57–64.

NL Agency. 2011. *Usability of Life Cycle Assessment for Cradle to Cradle Purposes—Position Paper*. Utrecht, The Netherlands: Ministry of Infrastructure and the Environment.

Norton, D.A., P.R. Espie, W. Murray, and J. Murray. 2006. "Influence of Pastoral Management on Plant Biodiversity in a Depleted Short Tussock Grassland, Mackenzie Basin." *New Zealand Journal of Ecology* 30, no. 3: 335–44.

OECD. 1991. *The Nature of Innovation and the Evolution of the Productive System, Technology and Productivity: The Challenge for Economic Policy*. Paris: OECD.

Orecchini, F. 2007. "A 'Measurable' Definition of Sustainable Development Based on Closed Cycles of Resources and Its Application to Energy Systems." *Sustainability Science* 2, no. 2: 245–52.

Prendeville, S., F. O'Connor, and L. Palmer. 2011. *Barriers and Benefits to Ecodesign: A Case Study of Tool Use in an SME*. IEEE International Symposium on Sustainable Systems and Technology (ISSST), 16–18 May, Chicago. IEEE.

Ramus, C.A. 2001. "Organizational Support for Employees: Encouraging Creative Ideas for Environmental Sustainability." *California Management Review* 43, no. 3: 85–105.

Reay, S.D., J.P. McCool, and A. Withell. 2011. "Exploring the Feasibility of Cradle to Cradle (Product) Design: Perspective from New Zealand Scientists." *Journal of Sustainable Development* 4, no. 1: 36–44.

Reid, A., and M. Miedzinski. 2008. "Eco-innovation: Final Report for Sectoral Innovation Watch." www.technopolis-group.com/resources/downloads/661_report_final.pdf

Reijnders, L. 2008. "Are Emissions or Wastes Consisting of Biological Nutrients Good or Healthy?" *Journal of Cleaner Production* 16, no. 10: 1138–41.

Rennings, K. 2000. "Redefining Innovation: Eco-innovation Research and the Contribution from Ecological Economics." *Ecological Economics* 32, no. 2: 319–32.

Rossi, M., S. Charon, G. Wing, and J. Ewell. 2006. "Design for the Next Generation: Incorporating Cradle-to-Cradle Design into Herman Miller Products." *Journal of Industrial Ecology* 10, no. 4: 193–210.

Sarkar, S. 2007. *Innovation, Market Archetypes and Outcome: An Integrated Framework*. Heidelberg and New York: Physica-Verlag.

Schilling, M.A. 2008. *Strategic Management of Technological Innovation*. 2nd ed. New York: McGraw-Hill.

Song, Y.S., J.R. Youn, and T.G. Gutowski. 2009. "Life Cycle Energy Analysis of Fiber-reinforced Composites." *Composites Part A* 40, no. 8: 1257–65.

Spicer, A.J., and M.R. Johnson. 2004. "Third-party Demanufacturing as a Solution for Extended Producer Responsibility." *Journal of Cleaner Production* 12, no. 1: 37–45.

Stegall, N. 2006. "Designing for Sustainability: A Philosophy for Ecologically Intentional Design." *Design Issues* 22, no. 2: 56–63.

Stevens, G.A., and J. Burley. 1997. "3,000 Raw Ideas = 1 Commercial Success!" *Research Technology Management* 40, no. 3: 16–27.

Tinnesand, B. 1973. "Towards a General Theory of Innovation." PhD diss., University of Wisconsin, Madison.

Vollenbroek, F.A. 2002. "Sustainable Development and the Challenge of Innovation." *Journal of Cleaner Production* 10, no. 3: 215–23.

Wang, G., and R. Côté. 2011. "Integrating Eco-efficiency and Eco-effectiveness into the Design of Sustainable Industrial Systems in China." *International Journal of Sustainable Development & World Ecology* 18, no. 1: 65–77.

Zlotin, B., A. Zusman, and L. Smith. 2002. *Futuring the Next Industrial Revolution. ASQ's 56th Annual Quality Congress and Exposition, 20–22 May, Denver*. American Society for Quality Control.

CHAPTER 3

The Next Industrial Revolution

Salah M. El-Haggar

"First they ignore you, then they laugh at you, then they fight you, then you win."

—Mahatma Gandhi

Introduction

Before the industrial revolution, the primary sources of energy for civilizations were human and animal muscle power, which were supplemented with wind, wood, and water power. As the industrial revolution advanced, science and technology discovered new sources of energy including coal, oil, and gas, and later nuclear fission, that multiply by thousands and then by millions of times the amount of work that human labor could accomplish.

For a very long period spanning thousands of years, prior to the industrial epoch of the past 250 years, industrial changes were relatively slow, although the impacts could be far-reaching because societies relied on wave and wind power, and solar energy, for the production of food and goods (McLamb, 2011).

During the eighteenth century, the source of power was the most important factor in the location of industrial activity. Initially the location was determined by the availability of wind or water power. The manipulation of water in order to drive water mills became very important in the United Kingdom, where rivers were diverted and reservoirs built in the hills to supply mills. The greater the catchment area and fall, the more powerful and continuous the source of water power was likely to be. Effective water management of rivers and canals was a valued skill and resulted in relatively high capital investment in mill buildings and machinery, and infrastructural investment in canals. One of the best-known and grandest schemes is the mill at New Lanark in Scotland.

The development of water power and its role in the rise of the factory system of production are very important in the history of British industrialization, yet they are secondary compared to the replacement of charcoal, a product derived from wood, by coal. The development of coal mining and the use of steam power generated from coal is without doubt the central, binding narrative of the nineteenth century. But we must realize that the use of horse and water power remained important well into the early twentieth century. However, the trend was set and soon the environment felt the full impact of industrialization in the form of air and water pollution (McLamb, 2011).

The First Industrial Revolution

The first industrial revolution began in England in the period from 1750 to 1850. In 1781, James Watt invented the steam engine, which had direct impacts on agriculture, manufacturing, mining, transportation, and technology. These in turn had profound effects on the social, economic, and cultural conditions of the times. The revolution then spread throughout Western Europe, North America, Japan, and eventually the rest of the world. By 1850, England had become an economic titan, supplying two-thirds of the world with cotton spun, dyed, and woven in the industrial centers of northern England. It proudly proclaimed itself to be the "Workshop of the World" until the end of the nineteenth century, when Germany, Japan, and the United States overtook it.

The industrial revolution itself was a shift from hand and home production to machine and factory production. It was important because of the inventions of spinning and weaving machines operated initially by water power, which was eventually replaced by steam. This helped increase America's growth. However, the industrial revolution truly changed American society and economy into a modern urban-industrial state. When America went to war with Great Britain in 1812, it became apparent that America needed a better transportation system and greater economic independence. Therefore, manufacturing began to expand. Industrialization in America involved three important developments. First, transportation was expanded. Second, electricity was effectively harnessed. Third, improvements such as refinement and acceleration were made to industrial processes. The government helped protect American manufacturers by passing a protective tariff (Kelly, 2012).

Examples of First Industrial Revolution Innovations
Textiles
In 1794, Eli Whitney invented the cotton gin, which made the separation of cotton seeds from fiber much faster. The South increased its cotton supply, sending raw cotton north to be used in the manufacture of cloth. Francis C. Lowell boosted efficiency by bringing spinning and weaving processes together into one factory, leading to the development of the textile industry throughout New England. In 1846, Elias Howe created the sewing machine, which revolutionized the manufacture of clothing. Suddenly clothing began to be made in factories instead of at home (Kelly, 2012).

Cotton spinning used Richard Arkwright's water frame, James Hargreaves' spinning jenny, and Samuel Crompton's spinning mule (a combination of the spinning jenny and the water frame). The latter was patented in 1769 and came out of patent in 1783. The end of the patent was rapidly followed by the erection of many cotton mills. Similar technology was subsequently applied to spinning worsted yarn for various textiles and flax for linen.

Steam power
The improved steam engine invented by James Watt and patented in 1775 was at first mainly used to power pumps for pumping water out of mines, but from the 1780s was applied to power other types of machines. This enabled rapid development of efficient semi-automated factories on a previously unimaginable scale in places where water power was not available. For the first time in history people did not have to rely on human or animal muscle, wind, or water for power. The steam engine was used to pump water from coal mines, to lift trucks of coal to the surface, to blow air into iron furnaces, to grind clay for pottery, and to power new factories of all kinds. For over a hundred years the steam engine was the king of industry.

Communication
As the United States expanded, better communication networks became increasingly important. In 1844, Samuel F.B. Morse created the telegraph, and by 1860, this network ranged throughout the East Coast to the Mississippi River (Kelly, 2012).

Transportation

The Cumberland Road, the first national road, was begun in 1811. (It eventually became part of Interstate 40). River transportation was made more efficient with the construction of the first steamboat, the *Clermont*, by Robert Fulton, made possible by James Watt's invention of the first reliable steam engine.

The opening of the Erie Canal created a route from the Atlantic Ocean to the Great Lakes, thereby helping stimulate the economy of New York and making New York City a great trading center.

Railroads were of supreme importance to the increase in trade throughout the United States. In fact, by the start of the Civil War, railroads linked the most important midwestern cities with the Atlantic coast. Railroads further opened the west and connected raw materials to factories and markets. A transcontinental railroad was completed in 1869 at Promontory, Utah.

Building on the great advances of the industrial revolution, inventors continued to work throughout the rest of the nineteenth and early twentieth centuries on ways to make life easier while increasing productivity. The foundations laid throughout the mid-1800s set the stage for inventions such as the light bulb (Thomas Edison), the telephone (Alexander Bell), and the automobile (Karl Benz). Further, Henry Ford's creation of the assembly line, which made manufacturing more efficient, helped transform America into a modern industrialized nation. The impact of these and other inventions of the time cannot be underestimated (Kelly, 2012).

Conventional Capitalism

According to Hawken, Lovins, and Lovins in their book *Natural Capitalism: Creating the Next Industrial Revolution* (2010), economic progress flourishes best in free-market systems of production and distribution, because when profits are reinvested, the productivity of labor and capital is increased. Large-scale manufacturing is encouraged and growth in total output (GDP) maximizes human benefit. Deficits in available resources will encourage the development of substitutes, and free-enterprise and market forces will optimize the use of labor and resources. However, a compromise between economic growth and a healthy environment must be made to ensure a high standard of living.

The industrial boom that occurred as a result of the industrial revolution changed the means of production and distribution of goods, especially after the introduction of machines powered by water, wood, charcoal, coal, oil, and eventually electricity. This accelerated and condensed the work previously done by labor. A single spinner in 1800, for example, could do the work of two hundred workers in 1700. Manufacturing rose to the level of large-scale production at reduced costs, which improved the standard of living and increased the demand for products in many industries. It also increased the demand for transportation, housing, education, clothing, and other goods, creating the foundation of modern commerce.

The present capitalist industry model is based on the fallacy that natural and human capital have little value as compared to the final output. In the new world of capitalism and industry, wealth was no longer based solely on land, animals, labor, and gold. Instead, more and more, it was based on innovation, on doing more with less, on science and technology, and on free trade between innovators. With the progress in science and technology in the nineteenth and twentieth centuries, physical and human resources are utilized much more efficiently, producing hundreds and thousands of times more wealth than ever before.

Adverse Consequences of the First Industrial Revolution

With the rapid increase in human population, a new pattern of scarcity emerged. Before the industrial revolution, production depended on mainly on labor, while natural capital was abundant and unexploited. The situation was reversed after two centuries of industrial progress, as labor became abundant and natural resources were exploited in an unsustainable matter to the extent that they became scarce.

Applying the same economic logic that drove the industrial revolution to this newly emerging pattern of scarcity implies that if there is to be prosperity in the future, society must make its use of resources vastly more productive, gaining multiple benefits from each unit of energy, water, materials, or anything else borrowed from the planet and consumed. Achieving this degree of efficiency may not be as difficult as it might seem because, from a materials and energy perspective, the economy is massively inefficient (Hawken, Lovins, and Lovins, 2010).

The resulting ecological strains are also causing or exacerbating many forms of social distress and conflict. For example, grinding poverty, hunger, malnutrition, and rampant disease affect one-third of the world and are growing in absolute numbers; not surprisingly, crime, corruption, lawlessness, and anarchy are also on the rise. The number of refugees has increased throughout the world. Many people around the globe are unemployed, or work low-paying jobs that do not allow them to support their families (Hawken, Lovins, and Lovins, 2010).

Human and natural resources have been regarded not as free and inexhaustible, but rather as finite, valuable factors of production.

Science and technology parks (see chapter 10) are an evolution of industrial concentrations started in the United Kingdom and the United States in the wake of the first industrial revolution. The concept of industrial concentration was quickly adopted. The first industrial estate in the United States was the central manufacturing estate in Chicago, developed in 1905. The idea of concentrating companies in one single area became increasingly important shortly after the Second World War because science had made vital contributions to victory (such as atomic energy and radar). After the first industrial revolution, which resulted in prosperity and mass production, its negative side effects appeared in natural resource losses such as trees and minerals, which are not renewable. In addition, it caused air, water, and land pollution. The first industrial revolution also increased greenhouse gases that raised temperatures. Gradually global warming began to affect millions of lives. In short, the first industrial revolution resulted in the long term not only in unsustainable production and consumption, but in huge amounts of solid and liquid waste that cause pollution, consume even more natural resources, and affect land-use plans.

The industrial revolution marked a major turning point in earth's ecology and humans' relationship with their environment, dramatically changing every aspect of human life and lifestyles. However, it did not become an unstoppable juggernaut overnight. The full economic effects of the industrial revolution were not realized until about a hundred years into it, in the 1800s, when the use of machines to replace human labor spread throughout Europe and North America. This transformation is referred to as the industrialization of the world. These processes gave rise to sweeping increases in production capacity and would eventually affect all basic human needs, including food production, medicine, housing, and clothing. Not only did society develop the ability to have more things faster, it was also able to develop better things. The full impact of this process on the world's psyche would not begin to register until the early 1960s, some two hundred years after its beginnings. From human development, health, and longevity, to social improvements and the impact on natural resources, public health, energy usage, and sanitation, the effects have been profound. These industrialization processes continue today (McLamb, 2011).

Sustainable/Green Economy

There is now a need for a fundamental conceptual shift away from the design of the current industrial system according to the 'cradle-to-grave' (C2G) approach, which generates toxic material flow, toward a 'cradle-to-cradle' (C2C) system in which renewable material and energy are generated through a closed loop of waste reusage in a safe, innovative, cost-reducing, and environmentally friendly manner.

The cradle-to-cradle system is a science- and values-based vision of sustainability that articulates a positive, long-term goal for engineers. It is defined as the design of a commercially productive, socially beneficial, and ecologically intelligent industrial system, based on optimization techniques.

The C2C approach, considered as a long-term investment, has many benefits for both the environment and the economy. It optimizes time, money, and resources in a way that not only does not harm the environment, but also benefits the economy, encouraging not only sustainable economy but also new perspective and practices for engineering and design ('green design') (McDonough and Braungart, 2002).

Sustainable economy is an economic system that meets the current needs of its members without compromising the prospects of future generations. Some of the initial responses for achieving sustainable economy include the 'circular economy' (in China) and the 'recycling economy' (in Japan and Germany), eco-industrial parks and networks, closed loop, and the C2C approach. Environmental sustainability should be taken into account in order to minimize adverse impacts for all major economic sectors, not just heavy industries.

Globally there is increasing environmental pressure on basic resources including land, air, and water, and competitive demand for oil, steel, copper, alumina, cement, and other basic commodities, industrial products, and construction. This is driven not only by new development but also by deteriorating urban, rural, and industrial infrastructure, as well as commercial and residential communities.

Sustainable development is a requirement in both developed and developing countries to efficiently use available resources. These resources are now being consumed at a level exceeding the improvements in manufacturing productivity in the areas of land-use planning and development, transportation, design and construction of the built environment, and many others.

Innovation is crucial for sustainable economic growth, as discussed in chapter 1. To ensure that innovation gives attention to all components of the ecosystem, a comprehensive National Innovation Strategy is required to stimulate public, private, and university dialogue around specific measures to take full advantage of the national potential that will result in economic growth. Sustainable economy is the result of the second industrial revolution.

The Second Industrial Revolution

The first industrial revolution was not well designed and planned; it progressed through the years based on immediate demands and needs. It is very important to have the second industrial revolution well prepared and planned in order to avoid the adverse impacts of the first industrial revolution:

- Similar to its predecessor, the first industrial revolution, it will swiftly change the world, as the population will realize that wastes can be regarded as resources. As the world is approaching a time of scarce resources and ever-increasing oil prices, on which the whole world's industry heavily relies, the situation threatens to return to pre-revolution times where the people had to employ their own muscles in simple daily work in order to live. Productivity can take a stunning climb if wastes are utilized as resources. Just as in the first industrial revolution, it could open up the world's capacity for inventions. The change will be so powerful and influential socioeconomically, on the level of the individual citizen as well as on the country level, that it deserves to be called a revolution. It will not come to an end. It will be sustainable. It will broaden people's capacity for future innovations. It will be an industrial evolution.
- The problem will be even larger if we allow ourselves to revert to pre-revolution practices.
- The C2C concept of waste appreciation needs to be adopted throughout the population.
- The second industrial revolution will build on the first industrial revolution. There is already a boom in commerce and channels of communication and transportation. There is already a reorganization of production. Technology continues to evolve. The second industrial revolution will improve people's lives with minimum adverse effects on the environment.
- The socioeconomic effect of the second industrial revolution, in terms of health and standard of living, will be positive.

We are currently depending on very limited resources. The first industrial revolution has driven the world's population to a paradigm of production at the expense of the human condition.

The capitalism of a new industrial revolution is based on a different mindset and values than the conventional capitalism (Hawken, Lovins, and Lovins, 2010). In this view, the environment is important, as it is the "envelope containing, provisioning and sustaining the entire economy." The availability of natural resources is the limiting factor to future economic development, especially the ones that have no substitutes and those that currently have no market value. Many factors contribute to the loss of natural capital, including badly designed business systems, population growth, and wasteful consumption. The new capitalism also promotes the idea that future economic progress can best take place in democratic, market-based systems of production and distribution in which all forms of capital are fully valued, including human, manufactured, financial, and natural capital. In addition, increasing resource productivity, and the quality and flow of services, are keys to the best use of human, financial, and environmental resources. Economic and environmental sustainability depends on redressing global inequities of income and material well-being. The best long-term environment for commerce is provided by true democratic systems of governance that are based on the needs of people rather than business.

The second industrial revolution must ensure the preservation of valuable social and natural processes to serve a growing population. This will be a critical need for coming generations.

Natural Capitalism

Capitalism, as currently practiced, is a financially profitable but non-sustainable approach in human development. The traditional definition of 'capital' is accumulated wealth in the form of investments, factories, and equipment. Natural capitalism, on the other hand, recognizes the critical interdependency between the production and the use of human-made capital and the maintenance and supply of natural capital. Natural capital refers to the earth's natural resources and the ecological systems that provide vital life-support services to society. It is being degraded by the very wasteful use of resources such as water, energy, and materials. Natural capitalism is a new business model that enables companies to fully realize different opportunities in markets, finance, and materials. The journey to

natural capitalism involves four major shifts (principles) in business practices, as discussed below (Hawken, Lovins, and Lovins, 2010).

Radical resource productivity

Exponentially increasing resource productivity is the cornerstone of
natural capitalism. Effective resource use has three significant benefits
that will result in lower costs for business and society, which no longer
has to pay for causes of ecosystem and social disturbance. This slows
resource depletion at one end of the value chain, lowers pollution at the
other end, and provides employment opportunities in useful jobs.

Biomimicry

Biomimicry is the concept of avoiding the generation of waste by
redesigning industrial systems to enable the constant reuse of materials
and continuously closed cycles, and often eliminating toxins.

Service and flow economy

This refers to the closed-loop cycle of material use and building a sustainably strong relationship between the producer and consumer. In
essence, the producer will acquire the product again from the consumer
at the end of its life cycle. Disposal of the waste becomes a shared
responsibility between the two parties.

The business model of traditional manufacturing rests on the sale of
goods. In the new business model, "service and flow" value is instead
delivered as a continuous flow of service—such as providing illumination rather than selling light bulbs. This aligns the interests of providers
and consumers in ways that reward them for resource productivity.

Investing in natural capital

This idea reverses the mindset of exploiting natural capital and struc-
turing it for industry investment, by investing in natural capital to
sustain, restore, and expand its stocks. This is crucial for a more abundant ecosystem and natural resources.

If these four changes are implemented, they will have positive
impacts on markets, finance, materials, distribution, and employment.
They will also reduce environmental degradation, perpetuate economic
growth, and significantly increase employment.

The current industries run on "life-support systems that require
enormous heat and pressure, are petrochemically dependent and

materials-intensive, and require large flow of toxic and hazardous chemicals" (Hawken, Lovins, and Lovins, 2010). These industrial "empty calories" end up as pollution, acid rain, and greenhouse gases, harming the environment and social and financial systems. But virgin resources will continue to be depleted, rather than waste resources reused, as long as the damage caused by their depletion is unaccounted for and their prices are maintained at deceptively low levels. People take it for granted that solid waste is disposed of in landfills, dumpsites, and waterways, and molecular waste finds its way to the atmosphere, soil, plants, surface water, groundwater, wildlife, and humans, as long as this process is partly or completely invisible.

Industrial processes generate waste in the form of ash, slurry, sludge, slag, flue gases, and construction debris. These wastes accumulate in nature and do not recycle naturally, unlike the biological processes that are regulated by such natural factors as seasons, weather, sun, soil, and temperature, all of which are governed by feedback loops. Feedback in nature is continual.

Since the first industrial revolution, unfortunately, most industries and world economies have largely ignored the environmental impact.

In society, waste takes a different form: human lives. According to the International Labor Organization in Geneva, nearly a billion people either cannot work or have marginal jobs that cannot support their families. In Europe, for example, unemployment has risen from 2% in the 1960s to 15% in 2009. In other parts of the world, it has reached as high as 40% (Hawken, Lovins, and Lovins, 2010).

The Third Industrial Revolution

Some authors believe the next industrial revolution will take one of three forms: a waste = food approach, sustainable economy, or C2C. The author would like to consider the next industrial revolution as a continuation of recent efforts to convert any type of waste into product according to the C2C concept in order to approach a sustainable economy. This will not only conserve natural resources but also protect the environment and provide job opportunities. The third industrial revolution will be based on strategies such as green industry, zero waste strategy, net zero energy strategy, net zero water strategy, zero pollution strategy, carbon neutral strategy, and low carbon strategy, where either no waste is generated or the waste of one industry is used to initiate new industries nearby (Nemerow, 1995).

Green industry

'Green' or 'sustainable' industry is a term used to refer to industrial businesses that produce environmentally friendly products or use industrial ecology in their operations. The green industry aims at eliminating or reducing emissions and industrial waste, as well as generating environmentally friendly products and services. By producing green products, green industries are contributing to improving the quality of the environment and conserving the natural resources used as raw materials. Green industries reduce pollution and waste by using waste and emission-free technologies and implementing efficient recycling processes (El-Haggar, 2009).

There has been increasing demand for green industries due to the major negative impact that traditional industries have on the environment. Traditional industries have a high rate of consumption of energy, water, and raw materials, in addition to gas emissions and waste production. According to the United Nations Industrial Development Organization (UNIDO), the manufacturing industry is known to consume 20% of global water use and most of the raw materials (Leuenberger, 2012). Unfortunately, many industries use more resources than needed for production. This happens due to the use of either outdated technologies or inefficient processes. Traditional industries also generate greenhouse gas emissions such as CO_2, methane, and nitrous oxides, which contribute to pollution and global warming. Clearly, current production technologies, processes, and systems are unsustainable.

In an attempt to materialize and encourage the adoption of green industry, UNIDO launched the "Green Industry Initiative" in 2008 (VanNest, 2012). Its main goal is to integrate environmental and social considerations into the operations of all business and manufacturing enterprises around the world. This is to be achieved through proper use of energy, water, and raw material resources, as well as the creation and application of new technologies and practices.

The green industry strategy aims to save natural resources and protect the environment through two fundamental components: greening current industries and creating new green industries (VanNest, 2012). Greening current industries is an opportunity for traditional businesses to change to environmentally sound businesses. A change in operations, technologies, processes, or end products is required. There are many methods by which this change can be achieved. One is the efficient use of resources, whereby the production processes are optimized to use less material in the product and packaging. In addition to saving our natural resources, this method will consequently lead to less waste, too. Another method is replacing non-renewable fuels such as coal, oil, or natural gas by renewable energy sources like solar, wind, or bioenergy. This will result in fewer or no harmful factory emissions. A third possibility is to phase out toxic substances from the manufacturing of products. There is already a trend in some major international companies like Johnson and Johnson, which announced in 2012 that it would remove carcinogenic chemicals from all its cosmetics and toiletries within three years (VanNest, 2012).

The second component in the green business strategy is creating green industries, which include new green businesses and expanding already existing ones. A green industry's core business processes, activities, and technologies comply with environmentally friendly rules and practices. Another category of green industries is one that develops and produces renewable energy machines or technologies (UNIDO, 2011).

Net zero green strategy

"Net zero green strategy" means that the annual consumption rate of resources should match their production rate. Following net zero green strategies will lead to improved financial, social, and environmental outcomes. It should thus be included in planning and directing the organization, community, or country toward achieving the common goal of a green economy, as discussed in chapter 1. Ideally, this will lead to 100% conservation of resources and zero depletion. As will be discussed later, the net zero strategy applies to many different areas including energy, water, carbon emissions, pollution, and so on, as well as sectors such as buildings, cities, and communities.

Toward net zero energy strategy

A net zero energy strategy requires producing onsite energy equal to or exceeding the energy consumed. In such a system, onsite energy is usually generated using renewable energy sources such as solar, wind, water, or bioenergy (Soltaniehha, 2012). An implicit goal is to reduce the energy demand so that it can be balanced by the energy produced, allowing the system to reach the zero energy targets. The zero energy strategies are dependent on technological advancements and innovative solutions that can make these strategies achievable. The following objectives will support the realization of a zero energy strategy (Kingery, 2011):

- Improve energy efficiency in new projects
- Include energy as a factor during planning and design phases
- Reduce energy consumption in current facilities
- Eliminate or reduce dependence on fossil fuels

The net zero energy strategy is implemented by many organizations and businesses worldwide, such as the US Army and the Beddington Zero (Fossil) Energy Development, or BedZED, in London. The latter was designed from scratch for the purpose of exploring the potential of creating a more sustainable lifestyle that simultaneously meets modern standards of comfort. The overall goal of the project was to reduce carbon emissions through both shrinking energy consumption and using alternate energy sources. Specifically, the project set out to reduce electricity and water consumption, as well as energy consumption needed for transportation and indoor heating and cooling.

Toward zero waste strategy

The "Sustainability in Action: Towards Zero Waste Strategy" (TZW) sets the direction for future reductions in waste generation and increases in resources (Sustainability Victoria, 2012). The main objectives of the TZW strategy are to:

- increase materials efficiency and minimize solid and liquid waste generation
- increase the sustainable recovery of materials for recycling and reprocessing
- reduce the environmentally damaging impacts of waste (Sustainability Victoria, 2012)

Within the TZW framework, actions and strategies are planned or have been developed for specific places and/or waste streams, including the Metropolitan Waste and Resource Recovery Strategic Plan (MWRRSP) for all metropolitan waste streams, and the Solid Industrial Waste Management Plan (SIWMP) for all solid industrial waste.

TZW aims at converting waste to usable resources, leaving nothing to be incinerated or landfilled. Although the concept of zero waste is a broad one that applies to all kinds of resources, including materials, energy, and human resources, it is most commonly used to refer to materials. Waste takes many forms: liquid, solid, hazardous, non-hazardous, and so on. Zero waste can be achieved through various techniques included in the 7-R rule discussed in detail in El-Haggar (2007) for all waste streams. The TZW strategy applies to all kinds of organizations, including businesses, industries, schools, universities, and homes. Waste is considered an inefficiency that can be eliminated by efficient techniques and practices leading to cost reduction. For example, Hewlett Packard in California reduced its waste by 95% and saved $870,564 in 1998 (Zero Waste, 2013). The TZW strategy has been adopted by numerous governments and organizations worldwide. Major automotive companies like General Motors, Toyota, and Subaru are producing landfill-free plants. Additionally, there are many organizations encouraging the zero waste initiative, such as the Zero Waste Institute, Zero Waste Network, and Zero Waste Alliance. Furthermore, there have been several regional organizations in California, Scotland, and Europe that support this strategy.

Toward low-carbon/neutral-carbon strategy

Carbon emissions—carbon oxides, in particular—are known to have harmful effects on the environment, global warming, and human health. These pollutants are known to cause respiratory and heart diseases. Therefore, elimination or reduction of carbon emissions is urgent. The carbon emissions produced by different sources are shown in table 3.1 (Boardman, 2007). It is clear that traditional energy sources produce much higher carbon emissions than renewable energy sources.

Table 3.1. Carbon emissions from different sources

Fuel	gC/kWh
Grid electricity	115
Anthracite	86
Oil	72
Liquid petroleum gas (LPG)	64
Mains gas	52
Renewable energy	7

Low-carbon strategy aims at cutting carbon emissions. This can be achieved by increasing greenfield sites, reducing the use of private transport, using renewable energy sources, implementing new emissions-free technologies, and so on.

Neutral-carbon strategy is a similar concept. To be carbon neutral is equivalent to having a zero net carbon footprint, which means balancing the amount of carbon emissions into the atmosphere by an equal amount of offset that negates the negative carbon effect. Being carbon neutral is a process that involves calculating carbon emissions, trying to reduce or eliminate them (if possible), and balancing the remaining amounts. Balancing the emissions is achieved by "purchasing carbon offsets" (Bookhart, 2008), which can be done by planting green fields and trees or by investing in environmentally friendly technologies such as renewable energy production (solar or wind energy). One simple way to reduce carbon emissions is to turn off lighting, heating, and cooling services during non-working hours.

Many countries have adopted the zero carbon strategy. One example is the United Kingdom, which plans to reduce housing emissions by 80% by 2050. Additionally, it has set a zero-carbon goal in all new homes by 2016 (Bookhart, 2008).

Toward zero pollution strategy

The zero pollution strategy aims at eliminating all kinds of pollutants and contaminants from the environment. All categories of pollution, including air, water, and soil pollution, should be reduced to acceptable limits. The zero pollution strategy proposed by Nemerow (1995) recommends using environmentally balanced industrial complexes (EBIC) by recovering and reusing wastes onsite within the same facility to minimize waste as much as possible; recovering and selling wastes to other companies or organizations; and locating waste producers and users together in one industrial complex so that the wastes of one entity act as raw materials for another. Zero pollution strategy applies to all sectors using the C2C approach discussed in detail in chapters 1 and 2.

How the Third Industrial Revolution Evolves

Jeremy Rifkin describes how the current industrial revolution is drawing to a close, and why and how we should work to shape the next one. Rifkin, an economist and energy visionary, is a senior lecturer at the Wharton School's Executive Education Program at the University of Pennsylvania, the president of the Foundation on Economic Trends in Washington, D.C., the author of nineteen books, and an advisor to the European Union and to heads of state around the world. In his most recent treatise (2011), he says that the European Union set out two goals at the beginning of the current century: becoming a "sustainable, low carbon emission society" and enhancing the economy of Europe. The first goal requires the transformation from fossil fuel energy to renewable energy as recommended by the third industrial revolution. Changing the energy regime would allow for more complex economic relations. Rifkin also relates energy revolutions with communication revolutions, pointing out that in the nineteenth century, print technology became cheap when steam power was introduced to printing. The cost of printing decreased and the speed increased, making the printing process more efficient and available. This was a main contributor to the first industrial revolution. With the introduction of electricity in the twentieth century, telephone, radio, and television became the communication vehicles promoting the second industrial revolution, which depended on the oil-powered combustion engine, suburban construction, and mass production.

Energy historians deal only with energy, and communication historians deal only with communications, but in history, neither can be studied without the other. That is the framework that led Rifkin to this kind of search, and the third industrial revolution came out of that narrative on how history evolves.

What Is Meant by the Third Industrial Revolution

The third industrial revolution is based on a new convergence of communication and energy. The centralized electricity communication that characterized the twentieth century scales vertically. The Internet, by contrast, is a distributed and collaborative communication medium and scales laterally. The two technologies are now beginning to converge. A transition to distributed renewable energy sources is under way. These sources are distinguished from the traditional energies—coal, oil, gas, tar sands—that

> are found only in a few places, require significant military and geopolitical investments and massive finance capital, and have to be scaled top-down because they are so expensive. Those energies are clearly sunsetting as we enter the long endgame of the Second Industrial Revolution.

Distributed energies, by contrast, are found in some frequency or proportion in every inch of the world: the sun, the wind, the geothermal heat under the ground, biomass garbage, agricultural and forest waste, small hydro, ocean tides and waves. (Rifkin, 2011)

The Five Main Pillars of the Third Industrial Revolution

The twenty-seven member nations of the European Union have committed to establishing a five-pillar infrastructure for a third industrial revolution based on this new convergence of communication and energy. Rifkin was privileged to develop the plan that was formally endorsed by the European Parliament in 2007. The following description of the five pillars is cited from Rifkin 2011.

Pillar 1 is the fact that the EU set a goal of transforming 20% of its energy to renewable energy by 2020.

Pillar 2 illustrates the methodology of collecting distributed renewable energy. The initial plan was to harness solar energy from Spain, Greece, and Italy and transfer it via high-voltage lines to the rest of Europe. The same concept applies for wind energy from Ireland and hydropower from Norway. The idea then developed of having every one of the 191 million buildings in the EU act as "partial green power plants"—collecting solar energy on the roofs, wind energy off the sides of the building, geothermal energy from the ground below the building, and biomass energy from the conversion of garbage in the building. This approach will generate millions of jobs in the EU and create new opportunities for thousands of small and medium-sized enterprises over a 40-year period.

Pillar 3 deals with the storage of renewable energy. Solar, wind, and hydropower are intermittent sources of energy; thus, storage is crucial when 15–20% of the power grid depends on renewable energy sources. The EU has used flywheels, batteries, and water pumping as storage systems. It is also promoting the use of hydrogen as a storage medium for surplus electricity generated from photovoltaic panels on building roofs. Surplus electricity ionizes water to produce hydrogen, which is stored to be used later to generate electricity.

Pillar 4 deals with how the green energy would be shared. One proposal is to transform the power grid and electricity transmission lines into an energy Internet. Surplus electricity generated from the buildings will be sold back through this energy Internet.

Pillar 5 is about how to integrate transport into the third industrial revolution infrastructure. The buildings will be used to provide power to electric and hydrogen fuel cell vehicles. These vehicles can be plugged into the grid to obtain electricity, or sell it back to the grid depending on the price.

The five pillars must be connected in order to create the synergies that will form the infrastructure for a new economic paradigm in the twenty-first century.

How to Achieve the Third Industrial Revolution (Rifkin, 2011)

Fossil fuels have been influencing the progress of globalization, as the economy is closely linked to the price of oil. As it rises, so do the prices of all other goods and services, from the food that depends on petro-chemical fertilizers and pesticides to construction materials, pharmaceutical products, packaging materials, and clothes. Power, heat, light, and transport are also reliant on fossil fuels. As all of these prices rise, purchasing power will plunge, leading to a second collapse of the global economy.

The second significant obstacle is the Copenhagen climate change summit, where 192 countries came together to address the issue of global warming. Although scientists emphasized the dramatic shift in the climate of the planet due to the emissions of industrial-induced carbon dioxide, methane, and nitrous oxide, politicians did not give the issue enough attention. Governments are not appreciating the impacts on the earth's hydrological cycle and ecosystems, which are threatening a mass extinction of plant and animal life in the twenty-first century due to the rise in earth temperature. They are unwilling to take the necessary steps to combat climate change, putting in doubt the future viability of our species on Earth.

Rifkin assembled the CEOs and other senior executives from global companies and trade associations when oil peaked at $147 a barrel in July 2008. He announced that they needed "to create a sustainable new economic vision and a game plan to re-grow the economy based on bringing the Internet together with renewable energy" by working together to create the essential infrastructure for the third industrial revolution.

Rifkin also criticizes the utility companies who control power supply and transmission. In his opinion, their model is not functioning well due to the rise in fossil fuel prices, the effect this has on climate change, and the fact that the old centralized electricity grids are inefficient and increasingly dysfunctional.

Rifkin encourages the public to generate their own green energy. He foresees that the costs of these technologies will continue to drop, just as the costs of computers and information technology have dropped over the past thirty years. Generating electricity will become cheaper and more efficient, as feed-in tariffs will give early adopters a premium price for the green electricity they generate and sell back to the grid. He also predicts that in twenty years the technology for collecting and distributing energy is going to be virtually free, just like information.

Accordingly, the role of utility companies will change fundamentally: they will become managers of energy rather than producers, just as major computer companies like IBM replaced computer manufacturing with information management. The critical issue facing the power and utility companies is whether they have the expertise to manage the energy Internet to which the people generating their own green electricity are sending their surplus. They will have to set up partnerships with manufacturers to help manage their energy flows in their production processes, supply chains, and logistics networks.

The cost of energy is the factor that will determine which companies will be able to succeed in the transition between the second industrial revolution and the third. The role of the power and utility companies will be to help companies manage their energy flow, decrease their energy use, increase their productivity, and improve their profit margins. In return, the companies will share back the return on their energy savings from increased productivity with the power and utility companies as per a "shared savings" agreement.

The End of Mass Production

In his book *The New Industrial Revolution* (2012), Peter Marsh talks about the end of mass production. Marsh is a journalist who reports on developments in manufacturing-related industries for the *Financial Times*. He received the UK Business Journalist of the Year Award in the manufacturing category in 2002.

Marsh points out that, between 2000 and 2011, emerging economies like China have made advances, along with other poorer nations like India, Brazil, Eastern Europe, and Russia. Lower production costs, mainly linked to wages, encouraged the shift of manufacturing to these regions, even though the final consumers are in the west. Moreover, the growing demand for increasingly sophisticated factory-made products in the developing nations has encouraged more industrial investment in these locations (Marsh, 2012).

The share of manufacturing in the rich nations has dropped from 73% in 2000 to 54% in 2011. These countries, including North America, Western Europe, and Japan, are trying to make up for their losses in the share of production in the following ways:

- Changes in technology, such as 3-D printing, will generate new kinds of manufacturing operations.
- "Tailoring" of products to suit user requirements will require manufacturing to be located closer to product use so that the necessary design changes can be incorporated most effectively.
- The high costs of transportation and the environmental factors associated with long-distance transport of products are also factors.
- The rising expenses of manufacturing in emerging economies, due mainly to rising wages, is reducing the incentive of companies to produce in the developing countries. (Marsh, 2012)

Summary
The path to the third industrial revolution is to develop a net zero green strategy for energy, water, waste, and pollution, as well as encouraging low carbon/neutral carbon technologies. A new energy/industry/communication matrix will be needed, along with Rifkin's five-pillar infrastructure. This can be done by linking scientific and technological know-how and working with local, regional, and national governments and their respective business communities and civil society organizations to transform the infrastructure of the global economy and prepare the world for the next great economic era. The third industrial revolution will favor continental markets and continental political unions.

As an example, the European Union has recently entered into a partnership with the African Union to begin laying the infrastructure for a third industrial revolution, which will eventually join the two continents. For example, plans are being developed for a multibillion-dollar project, called Desertec, which will bring energy generated from solar and wind technologies from the Sahara Desert via interconnector cables to Europe, providing more than 15 percent of the European Union's total energy needs by 2050.

Questions

1. Compare the first industrial revolution and the third industrial revolution.
2. What is the motivation for the third industrial revolution?
3. Explain the concept of a net zero energy community, with examples.
4. How is Desertec linked to the third industrial revolution?
5. What are the impacts of Desertec on the global environment?
6. What is the carbon footprint for the industrial revolution? Use examples to quantify the carbon footprint.
7. Compare the economic crises (in different parts of the world) and sustainable economy.
8. Develop a net zero water strategy for a community. Define your community first.
9. Compare green industry and zero pollution strategy.

References

Boardman, B. 2007. "Home Truth: A Low-Carbon Strategy to Reduce UK Housing Emissions by 80% by 2050." University of Oxford's Environmental Change Institute. www.eci.ox.ac.uk/research/energy/downloads/boardman07-hometruths.pdf

Bookhart, D. 2008. "Strategies for Carbon Neutrality." *Sustainability: The Journal of Record* 1, no. 1: 34–40.

El-Haggar, S.M. 2009. "Industrial Solid Waste Utilization and Disposal." In *Environmental Engineering*, 6th ed., edited by Nelson Nemerow, Franklin J. Agardy, Patrick J. Sullivan, and Joseph Salvato. Hoboken, NJ: John Wiley & Sons.

———. 2007. *Sustainable Industrial Design and Waste Management: Cradle-to-Cradle for Sustainable Development*. Boston: Elsevier Academic Press.

Hawken, P., A. Lovins, and H.L. Lovins. 2010. *Natural Capitalism: Creating the Next Industrial Revolution*. New York: Little, Brown Company.

Kelly, M. 2012. "Overview of the Industrial Revolution." americanhistory.about.com/od/industrialrev/a/indrevoverview.htm

Kingery, K. 2011. "Army Net Zero Installation Initiative and Cost Benefit Analysis Activity." GreenGov Symposium. www.greengov2011.org/presentations/SustainablePlanning/GreenGov-2011-Sustainable-S1-KristineKingery.pdf

Leuenberger, H. 2012. "Introduction to Green Industry: UNIDO Executive Training for Government Officials from Viet Nam, United Nations Industrial Development Organization (UNIDO)." institute.unido.org/documents/ M3S1_ExecutiveTraining/executivetraining2/Presentations/Heinz%20Leue nberger%20-%20Green%20Industry%20Initiative.pdf

Marsh, P. 2012. *The New Industrial Revolution: Consumers, Globalization and the End of Mass Production*. New Haven and London: Yale University Press.

McDonough, W., and M. Braungart. 2002. *Cradle to Cradle: Remaking the Way We Make Things*. New York: North Point Press. www.mcdonough.com/writings/cradle_to_cradle-alt.htm

McLamb, E. 2011. "The Ecological Impact of the Industrial Revolution." *Ecology*, 18 September. www.ecology.com/2011/09/18/ecological-impact-industrial-revolution/

Mokyr, J. 1998. "The Second Industrial Revolution." faculty.wcas.north-western.edu/~jmokyr/castronovo.pdf

Nemerow, N.L. 1995. *Zero Pollution for Industry: Waste Minimization through Industrial Complexes*. New York: Wiley.

Rifkin, J. 2011. *The Third Industrial Revolution: How Lateral Power Is Transforming Energy, the Economy, and the World*. New York: Palgrave Macmillan.

Soltaniehha, M. 2012. "Passive and Net Zero Energy Residential Designs in a Cold Climate: A Simulation-based Design Process for the Next Generation of Green Homes." Fifth National Conference of IBPSA, USA.

Sustainability Victoria. 2012. "Review of Regional Waste Management Groups." Sustainability Victoria Urban Workshop. www.sustainability.vic.gov.au/ resources/documents/Future_Directions_Paper_Consultation_Draft.pdf

Thinkquest. 2012. "The Second Industrial Revolution." library.thinkquest.org/ C0116084/IR2.htm

UNIDO (United Nations Industrial Development Organization). 2011. "Introducing the Green Industry Platform." www.unido.org/fileadmin/media/ documents/pdf/Energy_Environment/Green_Industry/Green%20Industry%2 0Platform%20%20-%20%20Introductory%20Note.pdf

VanNest, H. 2012. "Johnson & Johnson to Phase Out Toxic Chemicals in Baby and Adult Products." CBS News. www.wtsp.com/news/local/story.aspx?storyid=268672

Zero Waste. 2013. "Case, 6 March." www.zerowaste.org/case.htm

CHAPTER 4

Innovation in the Plastic Industry 1: Upcycle of Plastic Waste: Plastic Rejects

Salah M. El-Haggar, Lama El Hatow, Yasser Ibrahim, and Mohamed Abu Khatwa

Introduction

A plastic is an organic material with the ability to flow into a desired shape when heat and pressure are applied to it and to retain the shape when they are withdrawn. Even though plastics have a variety of benefits, they are in fact detrimental to the environment. They consume a large amount of energy and materials, primarily fossil fuel that, when combusted, emits hazardous toxins into the air. Manufacturing of plastic also consumes a significant amount of water. Disposal of plastic bags also poses a concern. One study concluded that 1.8 tons of oil are saved for every ton of recycled polythene produced. The benefits of recycling plastics are numerous, and thus should be investigated and utilized to the fullest extent (El-Haggar, 2007).

Plastics can generally be classified into several categories (see fig. 4.1). Thermoplastics, such as black plastic bags, may be recycled on condition that they are not contaminated. If they are contaminated, then they are usually burned (causing harmful emissions) or landfilled, because it is too expensive to clean them for recycling. They are thus often referred to as "plastic rejects." Cleaning requires immense amounts of water, chemicals, and energy, which defeats the purpose of conservation. The challenge is to find a use for rejected plastic waste, even though the result will be a product with lower physical and mechanical properties than those of virgin plastic material. New material may be used as a basis of comparison, but the properties will differ tremendously.

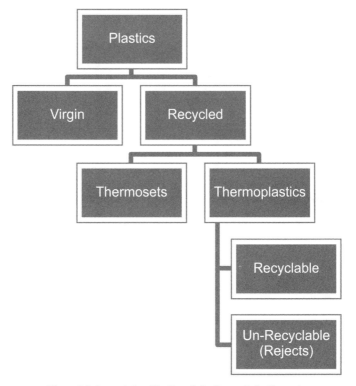

Figure 4.1. General classification of plastics and plastic waste

The main objective of this chapter is to highlight the utilization of these plastic rejects which are not recycled (a) because of the material structure, as in the case of thermosets, which don't melt once they are set and if they are burned they convert into ash, (b) because the rejects are made of mixed plastics with different properties and different characteristics, or (c) because they are contaminated and it is more expensive to clean and recycle them than to buy new raw materials, as in the case of black garbage bags (usually made of low-density polyethylene), plastic food packages, and so on. Types (b) and (c) are the most problematic, as thermoset wastes (type (a)) can be crushed into small particles and used as fillers and reinforcement in composites.

Although all plastics, such as polyethylene and polypropylene, are non-toxic, most plastics are not stable and are prone to degradation by UV radiation, heat, or mechanical pressure, where they are likely to release hazardous substances under the effect of these elements.

These plastic rejects produced from municipal solid or industrial/commercial waste can be used, after heating at low temperature, as a binder, where plastics of different compositions are mixed together with reinforcing materials such as regular sand, foundry sand, or glass cullets. One of its most important applications is to reduce manhole cover theft in developing countries by utilizing these materials in manufacturing manhole covers. No one will want to steal these covers, as they cannot be recycled, and they are immune to corrosion by sewage gases.

The environmental effects of plastics also differ according to the type and quantity of additives that have been used. Some flame retardants may pollute the environment (for example, bromine emissions). Pigments or colorants may contain heavy metals that are highly toxic to humans, such as chromium (Cr), copper (Cu), cobalt (Co), selenium (Se), lead (Pb), and cadmium (Cd), which are often used to produce brightly colored plastics. In most industrialized countries these pigments have been banned by law. The additives used as heat stabilizers (chemical compounds that raise the temperature at which decomposition occurs) frequently contain heavy metals such as barium (Ba), tin (Sn), Pb, and Cd.

The main advantages of recycling plastic rejects are:
- Using a natural resource that would otherwise be wasted
- Reducing or preventing solid waste going to landfill
- Reducing the costs involved in the disposal of solid waste, which ultimately leads to savings for the community
- Providing job opportunities
- Protecting natural resources and promoting sustainability
- Approaching a cradle-to-cradle production cycle

The disposal problem of plastic rejects has been growing globally in the last two decades, resulting in more incineration, landfilling, and open dumping in developing countries. Indeed, as the volume of plastic waste has increased, especially contaminated black plastic bags, legal restrictions have been imposed on those disposal methods, making recycling the only practical alternative. Although plastic recycling contributes to a significant reduction of the waste in need of final disposal, it results in the utilization of only about 30% of the total quantity of plastics; the remaining 70% are usually referred to as the rejects of the rejects. It has been estimated that those 70% account for about 5 million tons/year of non-recyclable mixed plastic waste in Egypt. Traditionally, those wastes have been dumped in open dump-sites in the deserts. They are often burned, producing harmful fumes, and the residues constitute a major source of soil and water contamination. These problems generated by the disposal of solid waste demonstrate the need for developing recycling techniques to achieve a reasonable level of hygiene, efficiency, and concurrence with international standards.

The potential use of plastic rejects in the production of composite materials for different applications in construction, sanitary, and mechanical applications, and in coastal zones, among others, is very important within the framework of the cradle-to-cradle concept and sustainable development.

Recycling Technology for Plastic Rejects

Recycling of plastic rejects was developed at the American University in Cairo (AUC) using three steps. First, innovate a new technology for un-recyclable materials like plastic rejects. Second, develop different economically valuable products from recycled plastic rejects according to demand and needs. The third step, market study, is a very important factor in order to guarantee the sustainability of the project. The manufacturing process of plastic rejects products consists of the following steps, as shown in figure 4.2:

- Sort plastic rejects from solid wastes
- Agglomerate/shred the plastic rejects (contaminated plastic waste)
- Mix the agglomerated/shredded plastic rejects with reinforced material and heat the mix indirectly to the required temperature
- Press the hot mix into molds using a hydraulic press to reach the required density and shape
- Cool the product

The hot mix can be easily molded into any form. The key issue behind this technology is the continuous mixing of the reinforcing material, such as sand and plastics, to guarantee homogeneous distribution of materials and good-quality products. Machinery required for this process is an agglomerator/shredder, an indirect heating furnace, a hydraulic press to form the products according to the mold shape, an overhead crane, and a conveyor-belt system.

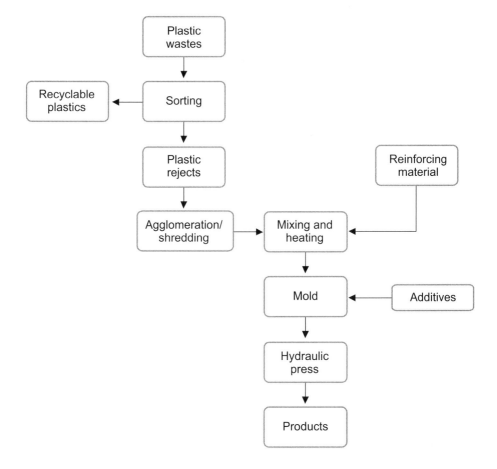

Figure 4.2. Schematic diagram of plastic rejects recycling technology

Manufacturing process

Plastic rejects are first shredded or agglomerated, then mixed with reinforcing material and heated indirectly to a temperature between 120°
and 180°C depending on the type of plastic, the quantity of materials,
and the product being produced. Thorough mixing is required in order
to obtain an even distribution of the mix within the product as a whole.
After heating for about 30–40 minutes, the produced material—which
is paste, not fluid—is then placed in a mold, where a hydraulic press will
form it into the required shape (fig. 4.3).

Figure 4.3. Manhole cover being shaped by hydraulic press

Size reduction techniques

There are several different size reduction techniques, such as cutting, shredding, and agglomeration of plastic rejects. Their main function is to increase the density of the material. One of the most important advantages of this 'densification' process is to minimize transport costs, as the volume of the material will be reduced. Furthermore, the small pieces generated from the densification process can be more easily fed into further reprocessing machines.

Cutting

The cutting process is the first stage in the process of plastic waste transformation. Its purpose is to break down plastic wastes into small pieces with low volume, so they can fit into the hopper of the shredder or agglomerator.

Shredding

The shredding process consists of feeding the cut plastic pieces into the shredder machine through a hopper. Inside the shredder machine, the cutting blades crumble the plastic wastes fed into it. The final products

Figure 4.4. Shredding machine

of the shredding are irregularly shaped pieces of plastic that can be used in the manufacturing process of the plastic reject technique. The shredding machine shown in figure 4.4 was used to shred laminated plastic bags (such as Chipsy bags) as one of the plastic rejects, to be easily folded into the mixture that will be fed into the furnace.

Agglomeration

Agglomeration is a pre-process for size reduction of plastics in the manufacturing process of plastic rejects. The agglomeration machine contains a cylinder with four rotating blades and another four stationary blades, as shown in figures 4.5 and 4.6. Both sets are used to cut the thin film plastic into small pieces. The effect of heat generated by friction during the agglomeration, along with small quantities of water, will convert the heated plastics into agglomerate. In other words, the agglomeration is the coalescing of small particles into a clump. The agglomeration is used to cut and pre-heat (or pre-plasticize) (fig. 4.7). Generally, the agglomeration process enhances the quality of the final product, as it increases the density of the material, which will result in a more continuous mixing of the material while heated. The bulk density of the raw materials in the agglomerator increases through shrinkage and partial plasticization. The material is cooled rapidly and solidifies as it is being cut, yielding coarse, irregular-shaped grains called crumbs. After the plastic waste is agglomerated, a door at the bottom is opened to release the agglomerated plastics, which are then placed in the mixing unit along with other reinforcing materials to produce the desired paste.

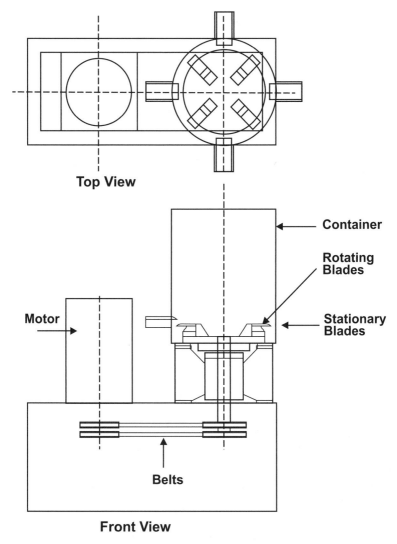

Figure 4.5. Schematic diagram of the agglomeration machine

Figure 4.6. Top view of the blades inside the agglomeration machine

Figure 4.7. Agglomerated plastics within the agglomeration machine

Feeding into the furnace

After the plastic rejects are cut, shredded, or agglomerated, they are mixed and heated with reinforced material such as sand in the heating furnace. The furnace consists of a combustion chamber that is used to distribute the heat uniformly along the cylindrical container. This furnace is equipped with a motor and gear driving system for the stirring process to blend the raw materials during heating. The mixture is heated for 30–40 minutes to attain the required temperature, and then the motor rotates the drum for stirring. The mixing process takes about 10–15 minutes. The paste is then dropped to be carried to the mold, where it will take the shape of the needed product and be compressed while cooling.

Indirect heating furnace

An indirect heating furnace is designed to blend the shredded plastic bags along with sand of a specified range of grain size and indirectly heat the mixture to a temperature between 120° and 180°C depending on the relative quantity of both materials, the type of plastics, and the needed product. Proper mixing is required for a better distribution of sand and plastic within the product as a whole.

The indirect heating furnace consists of a rotating drum equipped with an electric motor with a worm gearbox that transmits its power output to the gears located at the back of the rotating drum, as shown in figure 4.8. The gears rotate the drum, with the mixing shaft fixed on the carriage and rail, to stir the plastics, sand, and other additives into a homogeneous mixture. The fuel burner is located at the bottom of the rotating drum in a chamber to indirectly and efficiently heat the mix. The temperature inside the chamber is regulated through a thermocouple and a control panel. An electrical control panel is attached to the furnace to control the operation of the motor and automatically maintain a set temperature inside the machine's combustion chamber.

When the heating and mixing process is finished, the paste is extracted by moving the mixing shaft out of the drum, using the carriage and rails, as shown in figure 4.8. The collecting disk that is welded at the end of the shaft extracts the paste out of the drum, and drops it to be molded and pressed using the hydraulic press shown in figure 4.3.

The main functions of the heating furnace are the efficient heat distribution and isolation along the rotating drum, the proper feeding and extraction mechanisms, and the homogeneity of the paste.

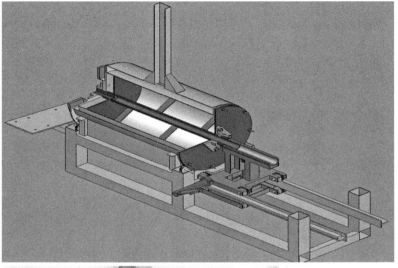

Figure 4.8. Mixing and heating furnace

Product Development from Plastic Rejects

The purpose of waste recycling is to develop and market the products that can be manufactured from it, in order to reach the goal of sustainability. This remanufacturing can create a number of economic opportunities. Just as market forces cannot be ignored when introducing a new product, they must also be taken into account when introducing remanufactured products. Product development from recycling plastic rejects is more challenging than the product development from recycling of waste or new materials because the properties of products from plastic rejects will change from one time to another due to their heterogenous nature. Similarly, the market analysis for these products may be more difficult than for new ones, because of consumers' concerns about quality, durability, and cost of maintenance. Therefore, continuous quality control is essential.

Construction Materials and Their Properties
Bricks
Bricks are the easiest products to manufacture out of rejects (figure 4.9). The major problem with these bricks, however, is adhesion. Because the bricks are made out of plastics and sand, the adhesive materials used with traditional bricks do not work. Adhesive materials that do work are very expensive, which adds to the cost of the brick. There is a cheap way to bind them together using pins for easy assembly and disassembly, as shown in figure 4.9, but the density of the bricks is very high, so they are much heavier than the ordinary variety. After the bricks were produced, it was discovered that the applications were not economical, as its price (including adhesives) was not competitive with that of ordinary bricks and adhesives. There was only one useful application for these bricks: in military purposes, because they are bullet-resistant.

Figure 4.9. Development of bricks

Interlocks

Interlocking bricks can be formed from plastic rejects by changing the shape of the mold. There are a number of applications for interlocks: pavements, gardens, factory floors, backyards, and so on. Interlocks are much more profitable than bricks, as they do not need adhesives. The appearance of the interlocks can be improved by coating the top layer with better-quality plastic waste. The cost of plastic rejects interlocks is 30–50% of traditional interlocks, depending on the top-layer coating. The top layer of plastic waste on the interlocks shown in figure 4.10 is a cheap coating made from plastic waste (clothes hangers that were crushed and melted in the heater, put in the mold, then pressed with the reject paste).

Plain bricks proved to be impractical due to the difficulty in adhesion because of water absorption. Interlocks, on the other hand, turned out to be very competitive. For the interlock the plastic coat was optimal because it is made out of clean waste plastic. This plastic coat covers the black color of the plastic rejects and has proved to be durable.

The brick/interlock material was composed of a mixture of plastic rejects and natural sand as a filler and reinforcing material. The plastic rejects included unsorted thermoplastic wastes from municipal solid

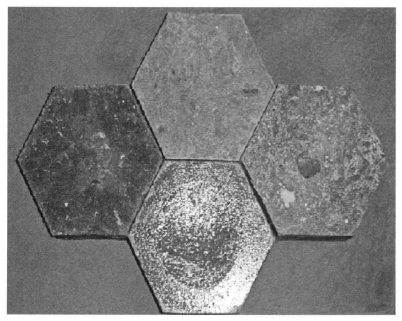

Figure 4.10. Development of interlocks

waste, the plastics recycling industry, the packaging industry, and other sources. The plastic rejects are a mixture of different types of plastics. The major waste constituents were low-density polyethylene (LDPE) and high-density polyethylene (HDPE), which together composed about 70% of the total waste. Small fractions of polystyrene (PS), polyethylene terephthalate (PET), and PVC were also present at a weight percentage of 20%. The balance consisted of plastic packaging waste with other minor waste constituents such as paper, wood, cardboard, small pieces of glass, food scraps, cloth fragments, and small metal chips. These plastic rejects are produced because municipal solid waste is not sorted at the source, which is typical in most developing and some developed countries. Inorganic siliceous sand with a fineness modulus (ASTM C136-96a, 1998) of 3.29 was used as filler in the bricks/interlock production. Two different sand particle sizes were employed, fine particles passing a 1.18mm (No. 16) ASTM sieve and coarser particles passing a 2.36mm (No. 8) ASTM sieve. The brick/interlock material was heated up while mixing, then poured into a steel mold and compacted using a hydraulic press.

A total of twelve different mixes were produced, incorporating different values for sand content, sand sieve size, and mixing temperature, as shown in table 4.1.

Table 4.1. Mix-design matrix for sample production
(Abou Khatwa, Salem, and El-Haggar, 2005)

Mix no.	Sand content (by weight)	Sand sieve size (ASTM)	Mixing temperature
Mix 1	20%	1.18mm	185°C
Mix 2	40%	1.18mm	185°C
Mix 3	60%	1.18mm	185°C
Mix 4	20%	2.36mm	185°C
Mix 5	40%	2.36mm	185°C
Mix 6	60%	2.36mm	185°C
Mix 7	20%	1.18mm	240°C
Mix 8	40%	1.18mm	240°C
Mix 9	60%	1.18mm	240°C
Mix 10	20%	2.36mm	240°C
Mix 11	40%	2.36mm	240°C
Mix 12	60%	2.36mm	240°C

There are no standards or codes to describe the testing procedures for recycled plastic reject composite material. Therefore, the standards for plastic testing, pedestrian and light traffic paving bricks, and cement tiles were used as guidelines to evaluate the properties of the composites produced.

Morphology

Morphological analysis (Abou Khatwa, Salem, and El-Haggar, 2005) conducted on samples of the recycled plastic rejects composite using stereomicroscope and scanning electron microscopy showed an intact morphology for the different sand concentrations, with the existence of high-volume fractions of voids of 6.3% for the 40% sand concentration, and 4.6% voids for the 60% sand concentration, but with larger sizes in the form of cavities for 60% sand. For the 20% sand concentration, the volume fraction of voids was 2.3%, which is less than that observed in the mixes prepared at a temperature of 185°C. However, 60% sand concentration prepared at 240°C produced a volume fraction of voids of 7%, which is higher than that observed in the mixes prepared at 185°C. (For more details see Abou Khatwa, Salem, and El-Haggar 2005). A comparison between samples prepared at 185°C and at 240°C revealed a better mix homogeneity for the higher mixing temperature. The dispersion of sand particles in the polymer matrix was enhanced and there were fewer areas of polymer segregation in the composite material. In addition, polymer segregation was much more pronounced in mixes containing high sand concentrations, as observed by the scanning electron micrographs of these segregated polymeric phases. It was also observed that weak interfacial bonds existed between the segregated areas and the polymeric matrix, as revealed by the inclusions present around the segregates. Incomplete melting of the recycled rejects was also observed, which resulted in lack of diffusion between the fragmented plastic wastes.

Mechanical Properties

Compressive strength

The compressive properties for all the mixes, as shown in table 4.2, were evaluated using the standard test method ASTM D 695M-91 (1998). The compressive strength values in this case can only be used as an indication of the stability of the material investigated.

The average compressive strength values ranged between 10.2 and 22.2 MPa at a strain of 0.1. Mix 3 (60% sand, sieve 1, and temperature

Table 4.2. Mechanical properties of prepared mixes
(Abou Khatwa, Salem, and El-Haggar, 2005)

Mix no.	Compressive strength (MPa)	Flexural strength (MPa)	Shore D hardness no.
Mix 1	10.154	12.241	58.28
Mix 2	11.897	11.184	63.37
Mix 3	15.622	10.447	66.2
Mix 4	12.821	15.191	59.43
Mix 5	15.502	11.307	64.17
Mix 6	16.478	10.725	68.53
Mix 7	10.992	16.463	60.4
Mix 8	13.125	21.239	67.73
Mix 9	20.25	11.26	68.8
Mix 10	13.122	13.55	61.88
Mix 11	17.981	16.345	69.12
Mix 12	22.247	13.936	69.87

185°C), mix 6 (60% sand, sieve 2, and temperature 185°C), mix 9 (60% sand, sieve 1, and temperature 240°C), and mix 12 (60% sand, sieve 2, and temperature 240°C) exhibited the highest compressive strength values, ranging from 15.6 to 22.2 MPa, while mix 1 (20% sand, sieve 1, and temperature 185°C), mix 4 (20% sand, sieve 2, and temperature 185°C), mix 7 (20% sand, sieve 1, and temperature 240°C), and mix 10 (20% sand, sieve 2, and temperature 240°C) showed the lowest compressive strength values, ranging from 10.2 to 13.1 MPa. Mix 2 (40% sand, sieve 1, and temperature 185°C), mix 5 (40% sand, sieve 2, and temperature 185°C), mix 8 (40% sand, sieve 1, and temperature 240°C), and mix 11 (40% sand, sieve 2, and temperature 240°C) had compressive strength values ranging from 11.9 to 18 MPa.

There is a direct proportionality between compressive strength value and sand content. This confirms what happens with filled elastomeric systems, where the filler particles reinforce the matrix by diverting the path of rupture, hence increasing the energy required to propagate a crack (Holliday, 1966).

There is also a rise in the compressive strength values associated with the increase in the size of sand particles. This suggests better dispersion and wetting characteristics associated with the large filler particles, leading to stronger interfacial bonds. It is known that for particle-filled composites, as the size of the filler decreases, their surface area increases, causing an increased particle–particle interaction leading to the formation of filler clusters (Nielson and Landel, 1994). Clusters represent weak points in the material, as they detach easily upon load application, creating voids and cavities. Moreover, particle clusters usually contain entrapped air, and it has been established that strength and modulus decrease with the increase in the amount of entrapped air (Nielson and Landel, 1994).

Following the ASTM specification C936-96 (ASTM Standards Source, 1998c), the average compressive strength for solid concrete interlocking paving units should not be less than 55 MPa for all test samples with no individual unit less than 50 MPa. However, it was shown during traffic tests in South Africa that the behavior of block pavements was unaffected by changes in compressive strength ranging from 25–55 MPa (Shackel, 1990). Hence, mix 12 prepared at a temperature of 240°C with the large sand particles (sieve 2) at 60% sand content has the potential for use as interlocking paving blocks. The same mix also satisfies the ASTM C902-95 specification (ASTM Standards Source, 1998b) for pedestrian and light traffic paving bricks, which requires a minimum compressive strength value of 20.7 MPa.

Flexural properties

The flexural strength for all the mixes was determined by test method I of the ASTM standard D790M-93 (1998). The results of all the flexure tests are summarized in table 4.2. The average modulus of rupture of the prepared mixes ranged from 10.4 to 21.2 MPa. The ASTM standard C410-60 (ASTM Standards Source, 1998a) for industrial floor bricks specifies a minimum modulus of rupture of 13.8 MPa for type M and L industrial floor bricks, and 6.9 MPa for type T and H. About half the investigated mixes satisfied the requirement for type M and L, and all the mixes satisfied the type T and H minimum value. In addition, the modulus of rupture of the produced composite material was higher than the modulus of rupture of clay bricks ranging from 3.5 to 10.5 MPa (Cordon, 1979). These results show the potential use of the composite material in industrial flooring applications, as well as structural applications.

The effects of sand content, sieve size, and mixing temperature on flexural strength were investigated. It is evident that for the low mixing temperature of 185°C, the flexural strength decreases with the increase in the sand content. On the other hand, for the high mixing temperature of 240°C, the flexural strength values increased with the increase in sand content from 20% to 40% with a maximum value for the 40% sand content of 21.2 MPa and 16.3 MPa for sieve size 1 and 2, respectively. However, increasing the sand content beyond this level resulted in a decrease in the flexural strength. This decrease in the flexural strength with the increase in sand content is attributed to the stress concentrations induced by the filler particles. This stress concentration will promote failure upon load application.

Hardness

The indentation hardness of the mixes was evaluated using the ASTM standard test method D2240-97 (1998). The use of this test method was intended only for comparison purposes and was selected due to its wide range of applications.

The results of the shore hardness tests are presented in table 4.2. The hardness values for all the investigated mixes ranged between 58 and 70. These numbers place the investigated composite material in the moderately hard plastics category with a shore D hardness ranging from 65 to 83 (Cordon, 1979).

The effect of sand content, sieve size, and mixing temperature on hardness was investigated. The hardness of the material increased with the increase in sand content, due to the increased hardness of the filler over that of the polymer matrix. Hardness also increased with the increase of the sand particle size from sieve 1 to sieve 2. However, the increase was limited, reaching a maximum of 3.5% between mix 3 (sieve 1) and mix 6 (sieve 2), both of which had 60% sand and a temperature of 185°C. The average increase in hardness values associated with increasing the sand particle size for mixes prepared at 185°C and 240°C was 2.25% and 2.02%, respectively. On the other hand, increasing the mixing temperature enhances the hardness significantly, due to improved adhesion between the polymeric matrix and the sand particles. The average increase for mixes prepared with sieve 2 was 4.82%, while mixes prepared with sieve 1 showed an average increase of 4.6%.

The optimum mix design for paving and tiling applications, based on the mechanical properties evaluated above, was found to be mix

12, prepared at 60% sand content using sieve 2 at a mixing temperature of 240°C. This mix gave the highest compressive strength and hardness values, as well as moderate flexural strength that remains higher than the requirements for clay and industrial floor bricks. Therefore, this mix will be the main focus for further evaluation of service properties.

Service Properties

Abrasion resistance

The abrasion resistance of bricks/interlocks was determined in accordance with the Egyptian standard test method ES 269-1974 (1974). The testing apparatus consisted of a rotating disc, a specimen-mounting plate holder, a hopper, and a distributor for feeding fresh abrasive. The loose abrasive utilized was quartz sand graded to pass a 0.9mm (No. 25) sieve and retained on a 0.6mm (No. 28) sieve.

The results for abrasion tests conducted on the group of samples prepared at 240°C using sieve 2 with different sand contents are shown in table 4.3. The average loss in thickness ranged from 0.05 to 0.24mm. Mix 10 (20% sand, sieve 2, and temperature 240°C) revealed the highest abrasion resistance value; this value decreased with increasing sand content.

Abrasive wear is the natural consequence of the shearing of junctions between surfaces caused by the introduction of abrasive particles from outside; by using soft plastics, the abrasive particles will sink below the surface, resulting in no further harm (Cordon, 1979). Thus, increasing the polymeric content on behalf of the filler reduces abrasive wear and seizure can be completely avoided. Moreover, when there is a weak adhesive bond between the filler and the polymeric matrix, fillers will increase the rate of wear of the material due to the ease with which the filler becomes separated from the matrix. Hence, increasing the sand content will decrease the resistance to abrasion.

Regardless of the differences encountered in the average thickness loss between the mixes, the highest value of 0.24mm for mix 12 (60% sand, sieve 2, and temperature 240°C) was 70% lower than the Egyptian standard limit for cement tiles, and 92% lower than the ASTM standard limit for solid concrete interlocking paving units. Hence, all mixes display excellent abrasion resistance (table 4.3).

Table 4.3. Abrasion resistance results of prepared mixes
(Abou Khatwa, Salem, and El-Haggar, 2005)

Mix no.	Sample no.	Weight before abrasion (g)	Weight after abrasion (g)	Bulk density (g/cm3)	Thickness loss (mm)	Average thickness loss (mm)
Mix 10	1	128	127	1.08	0.19	0.05
	2	164	164	1.08	0	
	3	156	156	1.08	0	
	4	122	122	1.08	0	
Mix 11	1	243	243	1.06	0	0.1
	2	183	182	1.06	0.2	
	3	177	176	1.06	0.19	
	4	179	179	1.06	0	
Mix 12	1	205	203	1.08	0.37	0.24
	2	196	195	1.08	0.19	
	3	209	208	1.08	0.19	
	4	217	216	1.08	0.19	

Density and water absorption

The density of all the prepared mixes was determined using procedure A of the ASTM standard test method D792-91 (1998). The relative rate of absorption of water for all mixes was evaluated by the twenty-four-hour immersion procedure of the ASTM standard test method D570-95 (1998) for water absorption of plastics.

The average density and water absorption results of the different mixes investigated are shown in table 4.4. All mixes resulted in low densities ranging from 1.12 to 1.68g/cm^3. The increase in density with the sand content is due to the high density of sand compared to polymeric material.

Table 4.5 lists the maximum water absorption percentages permitted by the ASTM and the Egyptian standards for various types of bricks, tiles, and paving units, compared to those produced by the composite material. It is evident from table 4.4 that even mix 1 (20% sand, sieve 1, and temperature 185°C), with the highest average water absorption value (1.6%) of all the investigated mixes, easily satisfies all the standard requirements for industrial floor bricks, interlocks, paving bricks, and

Table 4.4. Average densities and water absorption results of prepared mixes
(Abou Khatwa, Salem, and El-Haggar, 2005)

Mix no.	Density (g/cm3)	Water absorption (%)
Mix 1	1.12	1.6
Mix 2	1.27	0.71
Mix 3	1.63	1.51
Mix 4	1.13	0.7
Mix 5	1.29	0.94
Mix 6	1.59	1.1
Mix 7	1.12	1.04
Mix 8	1.39	0.64
Mix 9	1.59	1.29
Mix 10	1.16	0.98
Mix 11	1.41	0.83
Mix 12	1.68	0.85

Table 4.5. Maximum water absorption percentages for construction bricks and tiles
(Abou Khatwa, Salem, and El-Haggar, 2005)

Application	Standard specification	Water absorption (%)
Industrial floor brick	ASTM C410-60	1.5–12%
Solid concrete inter-locking paving units	ASTM C936-96	5%
Pedestrian and light traffic paving brick	ASTM C902-95	11–17%
Cement tiles	ES 269-1974	12%
Recycled plastic waste composite material		0.64–1.6%

cement tiles. It is evident from table 4.5 that the investigated plastic waste composite reveals superior water absorption resistance when compared to cementitious materials. However, the values are slightly higher than those expected for most plastics (less than 1% absorption). This could be related to the presence of imperfections in the prepared specimens, in the form of cavities, minor waste constituents such as paper, or pores resulting from the use of sand as filler material.

Resistance to chemical reagents

To assess the serviceability of the recycled plastic material when exposed to outdoor environment conditions, the ASTM standard practice D543-95 (1998) for evaluating the resistance of plastics to chemical reagents was employed. Three reagents were selected for the tests: benzene, sulfuric acid (3%), and sodium hydroxide solution (10%). Testing was conducted for 7, 14, and 28 days' immersion.

Tables 4.6, 4.7, and 4.8 list the average increase in weight over an immersion period of four weeks for the three reagents. It is apparent from the results that the materials possess excellent chemical resistance to acids and alkalis. The average increase in weight after 28 days was 1.37% and 3.38% for the sulfuric acid and the sodium hydroxide solution, respectively. However, the increase in weight after seven days accounted for nearly half the increase for the total immersion time. The

Table 4.6. Chemical resistance of samples subjected to 10% sodium hydroxide solution (Abou Khatwa, Salem, and El-Haggar, 2005)

Mix no.	Sample no.	Weight (g)				Increase in weight (%)			
		Initial	7 days	14 days	28 days	7 days	14 days	28 days	Average
12	1	39.08	39.66	39.93	40.14	1.48	2.18	2.71	1.95
	2	33.99	34.75	35.05	35.26	2.24	3.12	3.74	2.71
	3	37.32	38.11	38.38	38.70	2.15	2.84	3.70	3.38

Table 4.7. Chemical resistance of samples subjected to 3% sulfuric acid (Abou Khatwa, Salem, and El-Haggar, 2005)

Mix no.	Sample no.	Weight (g)				Increase in weight (%)			
		Initial	7 days	14 days	28 days	7 days	14 days	28 days	Average
12	1	30.34	30.58	30.73	30.74	0.79	1.29	1.32	0.81
	2	40.05	40.37	40.52	40.62	0.80	1.17	1.42	1.26
	3	37.90	38.22	38.40	38.42	0.84	1.32	1.37	1.37

Table 4.8. Chemical resistance of samples subjected to benzene
(Abou Khatwa, Salem, and El-Haggar, 2005)

Mix no.	Sample no.	Weight (g)				Increase in weight (%)			
		Initial	7 days	14 days	28 days	7 days	14 days	28 days	Average
12	1	30.46	32.37	33.48	34.06	6.27	9.91	11.81	5.59
	2	33.50	34.66	35.28	35.81	3.46	5.31	6.90	8.29
	3	32.83	35.14	36.00	36.39	7.04	9.66	10.84	9.85

increase in weight associated with the benzene reagent was much higher, reaching 5.59% after the first seven days and 9.85% after 28 days. It is well known that most polymers exhibit very high resistance to chemical attacks by acids and alkalis. However, they are less resistive to organic solvents, which react with the carbon atom chains and their pending side groups. This explains the increase in absorption rates for the benzene reagent. A visual analysis was also performed on the samples. It revealed no evidence of decomposition, discoloration, swelling, or cracking for the total immersion time.

Vicat softening temperature
The Vicat softening temperature was determined using the ASTM standard test method D1525-96 (1998), using three samples from the optimum mix 12. The results indicated a softening temperature between 93° and 110°C, which is much higher than the expected service temperature of 60°C.

Environmental-related Test
The health hazards of the investigated waste material were determined according to the water leaching test DIN 38414-S_4 (Kourany and El-Haggar, 2001). A fresh deionized water medium was used to determine the heavy metals content and leaching characteristics of the bricks. In this test, three specimens measuring 50x50x50mm from the selected mixes were immersed in 300ml of deionized water medium for 28 days. At the end of the immersion period, the media were analyzed for Cr, Cd, and Pb concentrations using a Perkin Elmer SIMAA 6000 spectrometer. A blank sample of the deionized water was also tested to determine the existence of any background levels of heavy metals.

   Table 4.9 lists the results of the leaching tests by water for 28 days, together with the World Health Organization (WHO) guideline values

(Kourany and El-Haggar, 2001) for heavy metal content in drinking water. The results indicate very low concentrations of heavy metals for the various mixes, as expected, since most of the wastes are plastic rejects with only minor potential sources of heavy metal contamination from the garbage dump sites. The samples are nearly free of Cd and contain very small amounts of Pb and Cr. The only exception was mix 12, which revealed high concentrations of 2.2 µg/L and 3.2 µg/L for Pb and Cr respectively. Nevertheless, all mixes showed concentrations far below the WHO guideline values, which proves that the recycled plastic waste composite material is safe from the health point of view.

Table 4.9. Heavy metal content by water leaching for 28 days
(Abou Khatwa, Salem, and El-Haggar, 2005)

Mix no.	Element concentration (µg/L)		
	Lead (Pb)	Chromium (Cr)	Cadmium (Cd)
Mix 4	0.1	0.2	0
Mix 5	1.4	1.5	0
Mix 6	0	0	0
Mix 10	0.4	0	0
Mix 11	0.1	0	0
Mix 12	2.2	3.2	0.1
Blank sample	0.2	0.8	0
WHO guidelines	10	50	3

Manhole Covers

A manhole or maintenance hole, sometimes called an inspection cover, is the top opening to an underground vault used to house an access point for making connections or performing maintenance on underground and buried public utility lines and other services, including sewers, telephone, electricity, storm drains, and gas. It is protected by a manhole cover designed to prevent accidental or unauthorized access to the manhole. Manhole base and cover are usually made out of cast iron or reinforced fiber plastic (RFP). They can also be produced from plastic rejects, as will be discussed later in this section.

Manhole covers usually weigh more than 100lb (50kg), partly so that the weight keeps them in place when traffic passes over them, and partly because they are often made out of cast iron, sometimes with infill of concrete. This makes them strong, but expensive and heavy. They usually feature pick holes into which a hook handle is inserted to lift them up.

Manhole covers are round because manholes are round, tubes being the strongest and most material-efficient shape against the compression of the earth around them. A circle is also the simplest shape that does not allow the lid to fall into the hole. Circular covers can be moved around by rolling, and they need not be aligned to put them back. However, the covers can be made in any shape that might be required by the geometry of the access opening.

Cast iron manhole covers

Manhole covers can take various shapes. Circular manholes range in diameter from 56cm to 150cm, and in weight from 50kg up to 136kg. They need to be heavy to prevent them from being lifted off the ground easily, especially since sewers can produce methane gas that can lift lightweight covers.

Manhole covers can be used outdoors (in streets and pavements) or indoors (in factories or other buildings). Following BS EN 124 standards, manhole covers are graded according to the maximum load they can withstand, and classified by type of service: the A range (15 kN) is for light service, B range (125 kN) is for medium service, C range (250 kN) is for hard service, and D range (400 kN) is for heavy-duty service (Clark-Drain, 2011).

Problems with cast iron manhole covers

Because of their value, cast iron manhole covers (CIMCs) are always a target for thieves. In Egypt, CIMC can cost around $150 apiece. Cities lose many covers to theft each year. China reported losing about 240,000 manhole covers in 2004 alone. According to the *International Herald Tribune*: "Since 2004, dozens of cities on every continent including Cardiff, Montreal, Milwaukee, Daegu, Chandigarh, and Johannesburg have experienced waves of manhole cover theft" (Reno, 2012). Thieves then sell them as scrap, according to *Newsweek*: "In 2001 scrap metal sold for $77 a ton. In 2004 it was $300 per ton, and today it's nearly $500 per ton." However, "they get only $10 to $20 per cover (Muir, 2012)." The problem is that after the theft the manholes are left open and uncovered (fig. 4.11), leading to terrible accidents, such as little children falling into

Figure 4.11. Open manhole after the cover was stolen

the open manholes and dying or sustaining severe injuries. The problem is a serious one in both India and China, causing at least eight deaths in 2004 (El-Haggar and El-Hatow, 2009a). Because CIMCs have such serious drawbacks, replacement materials became necessary.

Composite material manhole covers

Nowadays, some manhole covers are constructed of fiberglass-reinforced plastic (FRP) resin grate that fits into the standard manhole opening, as shown in figure 4.12. This type of material is strong enough to support a motor vehicle and light enough to be lifted out from below or above. Reduction in weight will decrease the potential for injury during opening and closure. It also has a unique anti-slip surface to prevent vehicles from skidding during wet or icy conditions. Another advantage of such manhole covers is that they can withstand adverse corrosive conditions. FRP manhole covers use a lamination compound craft on the surface layer between 5–8mm, where it joins the antioxidant to bear the marquls resin, and they are filled with an aggregate of quartz, corundum, and anti-friction materials, making the cover less

Figure 4.12. Manhole cover manufactured from a composite material (Enotes, 2009)

vulnerable to degradation and stirring phenomenon. Another advantage is that the way they are manufactured guarantees flush installation with the finished grade level to avoid trip hazard and snowplow damage. FRP manhole covers have shown during official testing that they can withstand more than 2 million fatigue shocks without cracking, giving them an extremely long service life. FRP manhole covers are highly heat resistant, even after more than three years of exposure in places where summer temperatures reach up to 43°C and road surface temperatures reach as high as 50–60°C. Manhole manufacturers recommend pursuing solidification craft-processing after the normal solidification process to increase the intensity of the product (fig. 4.12).

Materials used in manufacturing composite manhole covers
The materials used in the process of manufacturing composite manhole covers include rejects of thermoplastics, fiberglass, foundry sand, regular sand, and steel bars for reinforcement (El-Haggar and El-Hatow, 2009b). Regular sand used in the experimentation process was smaller than sieve size 8 (2.36mm).

Thermoplastic rejects

The plastic that was used was un-recyclable thermoplastic. All plastic items were placed together, generally consisting of contaminated garbage bags, thin film plastics, plastic bottles, and other such items. Contaminated plastic garbage bags should be cleaned and dried before agglomeration, a very costly and often unfeasible process. The contaminated plastic bags might be mixed with organic wastes and/or small pieces of glass and other impurities (which is why they are called 'plastic rejects'). A specific process is required in order to recycle these plastic rejects (Clark-Drain, 2011). First the plastic must be shredded or agglomerated into small granules. Once the agglomerates are produced, they are placed in the extruder (mixing unit) or heating furnace along with other reinforcing materials to achieve the desired paste.

Fiberglass

There are generally three main types of fiberglass: continuous, insulation wool, and special-purpose fibers. Continuous fiberglass is used in electrical insulation, and as a reinforcing agent for cement and plastics reinforcement, such as glass-reinforced concrete (GRC), glass-reinforced plastics (GRP), and glass-reinforced epoxy (GRE). It can generally be quite useful to reinforce plastics with fiberglass. Continuous glass fibers come in various forms, including continuous strand mats, chopped strands, milled fibers, and so on. Chopped strands are generally supplied in lengths varying from 1.5 to 50mm. This form of continuous fibers has been widely used in the reinforcement of thermoplastics. Milled fibers are generally smaller than 1.5mm and are powder-like. They are often used in electrical applications, but can also be used in the reinforcement of cementitious materials. Milled fibers provide stiffness and dimensional stability to plastics, but do not provide much strength (Muir, 2012).

Fiberglass is grouped into different classes. They include A glass (alkali-containing glass used for fiber manufacture), AR (alkali-resistant glass used for reinforcing cement), C (chemically resistant glass used for fiber manufacture), E (normal type of glass used in continuous glass fibers with a high electrical resistance and alkali content of less than 1%), and S glass (which has similar characteristics to E glass, but is stronger). These different classes are based on the manufacturing process of the glass fibers. Over 90% of glass fiber use falls under general-purpose products and are made of type E glass. The main benefit of glass fiber is its efficiency as a reinforcing agent in composite materials. In composite materials, approximately 0.5–2.0% of the weight of the product is composed of fiberglass particles (El-Hatow, 2006).

Glass fiber–reinforced plastics

Glass fiber–reinforced plastic (also known as GFRP or GRP) is a composite material using fiberglass and some form of plastic. The type of fiberglass used in GRP is often chopped strand mat (CSM), but woven fabrics and other forms are also used. Plastic resins are generally quite strong in compressive strength, but are relatively weak in tensile strength, while glass fibers are strong in tensile strength, but have no compressive strength. If the two are used together, they may thus resist both compressive and tensile loads. The tiny fiberglass strands prevent any cracks or flaws within the product from escalating and ultimately destroying the product. When a crack begins to grow within the specimen, its rate of growth will decrease considerably, and on occasion stop completely, because of the glass fibers in its path. Fiberglass used in the experimentation of this product is in the form of milled fibers of Type E glass.

Foundry sand

Foundry sand is a by-product waste sand from a metal foundry, or factory that produces castings of metal. It is used as a mold to cast the metals into the desired product. The sand mold is bonded together using organic binders. The sand casting system is the most commonly found system in the metal molding industry. The foundry sand is generally clean before it is used in the casting process, but once it has been used it may become hazardous. The used or spent material often contains residues of heavy metals and binders, which may become leachable contaminants. The residual sand is routinely screened and returned to the system for reuse. As the sands are repeatedly used, the particles eventually become too fine for the molding process and, combined with heat degradation from repeated pourings, require periodic replacement with fresh foundry sand (CWC, 1996). This 'spent sand' is black in color and contains a large amount of fines (particles of 100 sieve size or less), which may be a health hazard for humans in these industries. Most of this waste is landfilled, sometimes even serving as a cover material at landfill sites.

Steel bar reinforcement

A steel mesh is also incorporated into the manhole cover to enhance its mechanical properties. The steel mesh prepared as reinforcement constitutes 10% of the total volume of the cover. The cover is 60cm in diameter and 12cm thick, giving it a volume of 33,929cm³, of which 10% is 3,392cm³. The diameter of the steel bar is 12mm. Its spacing will vary, depending on the number of bars needed to constitute 10% reinforcement.

Sampling and testing manhole covers

The sampling began by organizing the different mixes and percentages of each component in each mix. There were six different types of mixes. The additional reinforcing elements were added in varying percentages to determine the most efficient combination. The percentages of each material are relative to the total weight of the mix. Each mix contained a certain quantity of plastic rejects, some form of sand reinforcement, and some designated percentage of steel mesh.

As table 4.10 shows, there were a total of six mixes. Each mix contained three replicates for accuracy in data, for a total of eighteen samples. The reinforcement percentage is a percentage of the total weight of the sample.

Table 4.10. Combination of mixes

Specimen no.	Components
1	20% foundry sand no steel
2	20% foundry sand steel Φ12
3	20% regular sand no steel
4	20% regular sand steel Φ12
5	30% regular sand no steel
6	fiberglass no steel

The reinforcing agents and designated mix are prepared as explained above. A steel mesh with a diameter of 12cm, placed one-third of the way from the bottom of the manhole cover mold, was used for mixes of specimens 2 and 4 to reinforce some of the covers, in order to determine the strength of the cover with and without the steel mesh. After this reinforcement, the mix was poured into the mold and compacted with a hydraulic press. It was left to harden for about 30 minutes, and then removed from the mold to continue drying and hardening. It was then tested for strength. A total of 24 different mixes of manhole covers were produced and tested. Figure 4.13 shows the schematic diagram of the entire process of producing a manhole cover.

The test that was performed on the manhole covers was a biaxial loading test, referred to as a bulge test. The test was done in accordance with ASTM Journal STP 563. It uses a hollow ring whose diameter is 2cm less than the opening in which the manhole cover sits, as shown in figure 4.14. A plunger with a diameter half that of the inner hollow ring (29cm) presses on the manhole cover from above. This entire apparatus is fixed onto a Universal Testing Machine (UTS) with a constant head rate of 5mm/min. Figure 4.15 shows the manhole cover being tested on the UTS machine. Figure 4.16 shows the manhole cover as a finished product before testing.

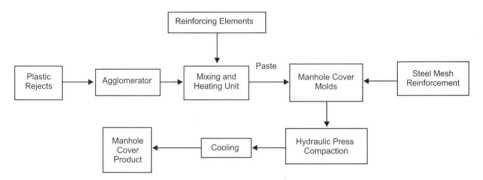

Figure 4.13. Schematic diagram of plastic rejects for manhole cover products

Figure 4.14. Hollow ring for bulge test

Figure 4.15. Manhole cover being tested on a UTS machine

Figure 4.16. Manhole cover as finished product

Results and discussion

It was found that foundry sand (20% reinforcement with $\Phi12$ steel) gave the most promising results. It appears that foundry sand is a stronger reinforcing material than regular sand. The composition of each of the manhole covers and the maximum allowable loads for each of them are shown in table 4.11.

As can be seen from table 4.11, $\Phi12$ steel with 20% foundry sand (specimen no. 2) yielded the strongest results, giving a maximum allowable load of 256 kN. The manhole covers that contained steel reinforcement were at least one-third stronger than those without steel reinforcement. This is a significant improvement in strength. (Strength also increases with the addition of sand.) It can therefore be concluded that steel reinforcement greatly enhances the strength and durability of manhole covers. Specimen 6 utilized fiberglass as a reinforcing element instead of steel. The average maximum load yielded by this specimen was 122 kN.

If we compare the BS EN 124 standards for manhole covers to the values obtained experimentally, it can be seen that the manhole covers produced from plastic rejects with different reinforcing elements come fairly close to the standards, and on occasion exceed them. Table 4.12 shows the maximum allowable load for manhole covers with a diameter of 30cm produced out of ductile iron material for the different grades of A, AA, AAA, B, C, and D according to BS EN 124. Grade A to AAA manhole covers are used in areas inaccessible to motor vehicles (pedestrian precincts). Grade B manhole covers are used in areas of only occasional vehicle access. Grade C is used in vehicular areas with medium to heavy traffic loads, and Grade D manhole covers are used in areas of fast-moving heavy vehicles.

Table 4.11. Average allowable loads for manhole covers (El-Haggar and El-Hatow, 2009b)

Specimen no.	Components	Max. load (kN)	Max. displacement (mm)
1	20% foundry sand—no steel	226	13.3
2	20% foundry sand—steel $\Phi12$	256	30
3	20% regular sand—no steel	114	15.8
4	20% regular sand—steel $\Phi12$	195	26
5	30% regular sand—no steel	188	17
6	Fiberglass—no steel	122	26.6

Table 4.12. Maximum load of ductile iron manhole covers with diameter of 30cm (BS EN 124, 1994)

Grade of manhole cover	Maximum load of ductile iron cover (kN)	Comments
Grade D	220	Heavy-duty traffic areas
Grade C	132	Medium to heavy vehicle traffic
Grade B	101	Occasional vehicle access pedestrian precinct parking areas
Grade AAA	52	Low vehicle access pedestrian precinct
Grade AA	31	Very light vehicle access pedestrian precinct (e.g., domestic driveways)
Grade A	8	No vehicle access pedestrian precinct (e.g., bicycles, pedestrians)

It can thus be seen that all compositions and combinations of the experimental plastic manhole covers can replace Grades A, AA, AAA, and B manhole covers of ductile iron. The majority of the compositions also rank higher than the Grade C specifications shown in table 12. Thus all the combinations used and manhole covers tested may be used in areas with occasional vehicle access and in pedestrian precincts. Two of the specimens are strong enough to replace Grade D ductile-iron manhole covers for heavy-duty traffic areas: 20% foundry sand with no steel and with Φ12mm steel mesh, which were able to withstand an average maximum load of 226 and 256 kN respectively.

Shoreline Erosion Protection Structures

Shoreline erosion protection structures vary in shape, function, and design. They include seawalls, bulkheads, revetments, breakwaters, and so on. Breakwaters are generally shore-parallel structures that reduce the amount of wave energy reaching a protected area.

Breakwaters are built to reduce wave action through a combination of reflection and dissipation of incoming wave energy. When used for harbors, breakwaters are constructed to create sufficiently calm waters for safe mooring and loading operations, handling of ships, and protection of harbor facilities. Breakwaters are also built to improve maneuvering conditions at harbor entrances and to help regulate sedimentation by directing currents and creating areas with differing levels of wave disturbance. Protection of water intakes for power stations and protection of coastlines and beaches against tsunami waves are other applications of breakwaters (Burcharth and Hughes, 2003).

When used for shore protection, breakwaters are built in near-shore waters and usually oriented parallel to the shore as detached breakwaters (fig. 4.17). The layout of breakwaters used to protect harbors is determined by the size and shape of the area to be protected as well as by the prevailing directions of storm waves, net direction of currents and littoral drift, and the maneuverability of the vessels using the harbor. Breakwaters protecting harbors and channel entrances can be either detached or shore-connected.

According to the US Naval Academy, "the cost of breakwaters increases dramatically with water depth and wave climate severity. Also, poor foundation conditions significantly increase costs. These three environmental factors heavily influence the design and positioning of the breakwaters and the harbor layout" (USNA, 2006).

According to Bakker et al. (2003), "Until World War II breakwater armoring was typically either made of rock or of parallel-epipedic concrete units [cubes]." Armor units are either randomly or uniformly placed in single or double layers. The governing stability factors are the units' own weight and their interlocking ability. Breakwaters are mostly designed with gentle slopes and relatively large armor units that are stabilized mainly by their own weight.

Figure 4.17. Breakwaters in Alexandria, Egypt

Figure 4.18. WhisprWave® HDPE
module (Whisprwave, 2005)

Recently, HDPE has been used in the manufacturing of breakwater structures, especially floating breakwaters. This technological advancement has been utilized by Wave Dispersion Technologies, Inc. (WDT), which developed the WhisprWave® floating articulated breakwater erosion/wave and wake control system to protect marinas, beaches, and private property subject to destructive or annoying wave/wake forces. The base building block of the WhisprWave® is its patented module, illustrated in figure 4.18. A standard module weighs approximately 36 lbs. empty, and can be filled with water to adjust buoyancy. The module can also be filled with marine-grade buoyant foam to be "puncture proofed" (Whisprwave, 2005).

Testing of erosion protection structures

Prior to selecting a sample size, the type of tests needed for the various shapes of breakwater structure/armor units were identified, to ensure that the samples produced would meet the minimum qualifications according to standards for specimen sizes. The general consensus was to choose either cube-shaped or rectangular-shaped (massive block) molds for ease of production and availability of manufacturing tools. The exact size was determined after a series of comparisons to identify the required tests for the sampled breakwater.

Breakwater armor units can be made of concrete, rock, or HDPE. However, the material utilized in these tests was plastic rejects mixed with sand, which had not been used before as an erosion protection structure or an armor unit, so no standard tests were available. Accordingly the testing for properties would be based on the experimenters' judgment of their suitability for the material used, based on standard tests for other materials. The tests that were conducted are shown in table 4.13.

Table 4.13. Breakwater erosion tests

Types of testing	Specific tests
Mechanical/physical property tests	Bulk specific gravity test (ASTM D792)
	Absorption test (ASTM D570)
	Compression test (ASTM D695)
	Abrasion test (ASTM C1138)
Accelerated weathering/chemical tests	Wet-dry test (ASTM D5313)
	Soundness test (ASTM D5240)
Environmental tests	Analysis of leachate from immersion in fresh and sea water

Process description

The primary considerations in the selection of a material for use in a coastal structure are availability, strength, durability, material life as compared to the desired life of the structure, costs, ease of maintenance, and maintenance costs. This section presents the results of the selected tests for the serviceability and suitability of plastic rejects as erosion protection structures.

The main materials used in physical/mechanical and accelerated weathering tests were solid waste plastic bags (LDPE). The bags were agglomerated to produce fine laminates as shown in figure 4.19. The

Figure 4.19. Agglomerated plastic bag

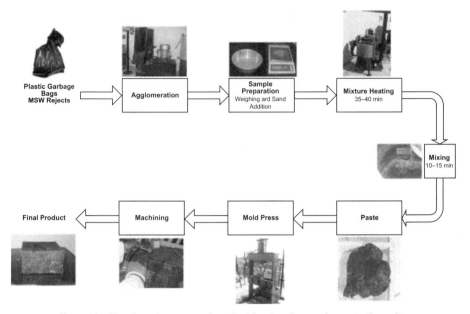

Figure 4.20. Manufacturing process flow chart for shoreline erosion protection units

agglomerated material was mixed with filler material sand in different ratios. The mix was then indirectly heated in a furnace at different temperatures to produce a paste, which was pressed in a prefabricated mold to the desired dimensions. The product was machined to cut off the excess paste. The final product was a shoreline erosion protection unit. A process flowchart is presented in figure 4.20.

Test procedures
Absorption test (ASTM D570)

'Absorption' refers to the amount of moisture absorbed. The moisture content of plastic is related to properties of the material, including electrical insulation resistance, dielectric losses, mechanical strength, appearance, and dimensions. Change in moisture content affects these properties. The water absorption rate depends on the type of exposure (immersion in water or exposure to high humidity), as well as the shape of the part and the inherent properties of the plastic.

In accordance with established testing standards, specimens of materials whose water-absorption value is potentially affected by temperatures near 110°C (230°F) were dried in an oven for 24 hours at

50 ± 3°C (122 ± 5.4°F), cooled, and immediately weighed to the nearest 0.01g. The conditioned specimens were then placed in a container of distilled water kept at a temperature of 23 ± 1°C (73.4 ± 1.8°F). The samples were entirely immersed. At the end of 24 hours, the specimens were removed from the water one at a time, all surface water was wiped off with a dry cloth, and they were weighed to the nearest 0.01g. Absorption percentage is calculated according to equation (1):

$$\text{Absorption \%} = \frac{(Wt_{wet} - Wt_{conditioned})}{Wt_{conditioned}} \times 100 \tag{1}$$

Specific gravity test

Specific gravity (relative density) is the ratio of the mass in air of a unit volume of the impermeable portion of the material at 23°C to the mass in air of equal density of an equal volume of gas-free distilled water at the same temperature. Density is the mass in air in kilograms per cubic meter of the impermeable portion of the material at 23°C.

The specimens were conditioned at 23 ± 2°C and 50 ± 5% relative humidity for 40 hours. The water temperature was measured, and the specimens were weighed in air to the nearest 0.01g. A special weight apparatus was arranged in which the sample was hung in a bucket of water hooked to a balance. The balance was reset to calculate only the weight of the sample in water. The specific gravity and density are calculated according to equations (2) and (3) below:

$$Specific\ gravity = \frac{Wt_{conditioned}}{(Wt_{conditioned} - Wt_{immersed})} \times 100 \tag{2}$$

$$Density = Sp\ Gr\ x\ 997.5 \tag{3}$$

Compression test

Compression tests provide information about the compressive properties of plastics, such as modulus of elasticity, yield stress, and compressive strength.

The specimens were conditioned at 23 ± 2°C and 50 ± 5% relative humidity for 40 hours in accordance with the test method ASTM D695. The 810 MTS testing machine with a load cell of 500 kN was used, and the machine was set at a compression rate of 1.3mm/min. The dimensions of all specimens were measured.

Wet–dry test

The wet–dry test is designed to simulate summertime conditions of alternating rainfall and subsequent drying by the summer sun. It also simulates the rise and fall of tidal movements and water levels in reservoirs, lakes, and rivers. Specimens are alternately soaked in water and heated for a specified number of cycles. The procedure as specified in Test Method D5313 is given for rock, so the preconditioning temperature was altered to suit the plastic material. The specimens were conditioned at $23 \pm 2°C$ and $50 \pm 5\%$ relative humidity for 40 hours and then weighed. Specimens were placed in a container with enough potable water that the specimens were fully immersed, and left to stand for a minimum of 12 hours. The samples were then dried in an oven at a temperature of 60 to 70°C (140 to 160°F) for a minimum of 6 hours. The cycle was repeated 5 times. To determine the percent loss, equation (4) is applied:

$$Percent\ loss = \frac{\left(Wt_{prior} - Wt_{post}\right)}{Wt_{prior}} \times 100 \qquad (4)$$

Soundness test

The soundness test (ASTM D5240) is an indirect attempt to simulate the expansion of water on freezing. The specimens are subjected to alternating cycles of immersion in a saturated solution of sodium sulfate or magnesium sulfate followed by oven drying to estimate qualitatively the durability of the substance under weathering conditions. This soundness test was also designed for rock, but it was used as a guideline to adapt this method for plastic rejects testing. However, the solution itself was prepared in accordance with the standard. Magnesium sulfate was selected because it is more destructive than sodium sulfate (Klieger and Lamond, 1994).

A saturated solution was prepared by dissolving equal grades of the salt in water at a temperature of 25 to 30°C (77 to 86°F). A measure of 350g of crystalline $MgSO_4$ salt was mixed per 1 liter of water, and 1240gm of $MgSO_4$ heptahydrate solution per liter of water (ASTM C88-99a, 1999). A solution mixture was prepared using 7kg of $MgSO_4$ heptahydrate; 500gm of crystalline salt were added to a total of 7 liters of water. The mixture was stirred at frequent intervals during the addition of the salt, and the solution was covered at all times to reduce evaporation and prevent contamination. The solution was allowed to cool to $21 \pm 1°C$ ($70 \pm 2°F$), stirred again, and was then kept at that temperature for 48 hours before use.

Barium chloride solution for washing was prepared by dissolving 5g of $BaCl_2$ in 100ml of distilled water. The specimens were conditioned at $23 \pm 2°C$ and $50 \pm 5\%$ relative humidity for 40 hours and then weighed. Specimens were immersed in the sulfate solution for not less than 16 hours nor more than 18 hours in such a manner that the solution covered them to a depth of at least 1.27cm (1/2in); however, the samples floated due to the increased salt concentration. The container was covered to reduce evaporation and prevent the accidental addition of extraneous substances. The samples were maintained at a temperature of $21 \pm 1°C$ ($70 \pm 2°F$) for the immersion period. Specimens were then removed and left to drain for about 15 ± 5 min, and placed in an oven. The samples were dried in the oven to a constant mass at $60 \pm 10°C$ at intervals of 2–4 hours. The process of immersion and drying was repeated for a total of five cycles. The specimens were then washed with barium chloride and hot water ($43 \pm 6°C$) and weighed. Calculation of the percent soundness loss is in accordance with equation (5) (ASTM D5240):

$$\% \ Soundness \ Loss = \frac{(Wt_{prior} - Wt_{post})}{Wt_{prior}} \times 100 \tag{5}$$

Abrasion test

This test determines the relative resistance of concrete to abrasion under water according to ASTM C1138-97. The test method was used as a guideline to simulate abrasion under water on the sampled plastic material. The specimen was placed in an agitation rotating device (washing machine) where 25 chrome steel grinding balls with a diameter of 15mm were laid down on the surface for the abrasion action. The standard test specifies a 1200 rpm rotating device to be operated for six 12-hour periods. The washing machine operated at 610 rpm, so a motor operating at 1500 rpm was fitted with agitating blades and attached to the assembly. The 12-hour period could not be achieved due to university constraints (i.e., no one was permitted to stay overnight in the lab), so the machine was left to operate for a 24-hour period and the specimens were then weighed. This process was repeated for a total of four

24-hour periods, equivalent to eight 12-hour periods. The abrasion loss is calculated according to equation (6):

$$Vt > \frac{(W_{air} - W_{water})}{G_w} \qquad (6)$$

Where:

Vt = Volume of specimen at desired time, m^3
W_{air} = Mass of specimen in air, Kg
W_{water} = Mass of specimen in water, Kg
G_w = unit weight of water, Kg/m^3 ($997\ Kg/m^3$)

To calculate volume lost at the end of any time increment:

$$VLt = Vi - Vt \qquad (7)$$

Where:

VLt = Volume of specimen lost by abrasion at end of time, m^3
Vi = Volume of specimen before testing, m^3
Vt = Volume of specimen at end of testing, m^3

To calculate average depth of wear:

$$ADAt = VLt/A \qquad (8)$$

Where:

$ADAt$ = Average depth of abrasion at end of test, m
VLt = Volume of specimen lost by abrasion at end of time, m^3
A = Area of top specimen, m^2

Leachate test

Two specimens from plastic rejects were immersed, one in sea water and the other in tap (fresh) water, for a period of 28 days. The water was then analyzed in accordance with DIN 38414-S4. The following parameters were analyzed:

1. pH
2. Total dissolved solids (TDS)
3. Total suspended solids (TSS)
4. Chemical oxygen demand (COD)
5. Nitrates and nitrites
6. Heavy metals (Cr, Cd, Pb)

Results and discussion

Absorption

The samples were checked every 24 hours during their 28-day immersion. Table 4.14 presents the data of absorption percent.

The absorption was observed to increase with the increase of sand content, which could be due to the rise in porosity of sand. However, the increase in temperature for the same sand content (40%) decreased the rate of absorption. This could be due to the fact that the plastic particles melt more and are thus more homogenously connected, reducing the interstitial spaces and allowing less absorption. The addition of water to the 30% sand content decreased the absorption rate. This could be due to more mixing of particles and filling of voids during sample preparation.

Table 4.14. Absorption percent averages for different mixes
(El-Haggar and Ibrahim, 2009)

Mix	Sand %	Temp.	% Absorption
0	0	140	0.1267
1	30	140	0.386
2	40	140	0.44
3	50	140	0.578
4	40	160	0.2504
5	40	180	0.2445
6	30, + 25% H_2O	160	0.1639

Specific gravity and density

Specific gravity and density are a measure of material occupation within a given volume. After the samples were immersed for 24 hours in the absorption test, they were conditioned and then tested as per ASTM D792. The results for specific gravity and density are given in table 4.15.

The 0% sand content samples floated, as they are less dense than water (average density of 0.982 gm/cm^3 less than that of water). Apart from flotation of 0% sand, the specific gravity was observed to increase with the increase in sand content.

Table 4.15. Specific gravity and density averages for different mixes
(El-Haggar and Ibrahim, 2009)

Mix	Specific gravity	Density (gm/cm³)
Mix 0	0.984	0.982
Mix 1	1.102	1.099
Mix 2	1.140	1.137
Mix 3	1.179	1.176
Mix 4	1.145	1.142
Mix 5	1.146	1.143
Mix 6	1.094	1.091

Strength and modulus of elasticity

The samples were compressed on the MTS 810 machine. Samples were subjected to loading at a rate of 1.3mm/min. Data for force and elongation were digitally measured and electronically stored, from which the stress–strain relationship was derived. The important properties measured in this experiment are the modulus of elasticity, which is a measure of stiffness, and the ultimate strength beyond which failure is initiated.

Under examination the 0% sand showed a higher ultimate strength, while the 50% sand resulted in the lowest strain value. In order to properly interpret the stress–strain diagrams, the Young's modulus (E) and the ultimate strength were calculated and identified for comparison purposes. Tables 4.16 and 4.17 present the data for Young's modulus (E) and ultimate strength using different sand contents and different temperatures.

Table 4.16. E and UTS with sand content variation at 140°C (El-Haggar and Ibrahim, 2009)

Sand content	Young's modulus (MPa)	Ultimate strength (MPa)
0%	283.911	22.216
30%	317.876	19.369
40%	324.168	20.481
50%	345.223	16.629

Table 4.17. E and UTS with temperature variation using 40% sand content samples
(El-Haggar and Ibrahim, 2009)

Temp (°C)	Young's modulus (MPa)	Ultimate strength (MPa)
140	324.168	20.481
160	293.747	16.801
180	287.109	14.209

The stiffness (E) increased with the increase in sand content, possibly due to the fact that the filler material is stiff, thus adding to the stiffness of the plastic rejects. However, when comparing the temperature effect on the 40% sand content sample, the stiffness was observed to decrease, and again this could be due to the homogenous distribution of plastic particles. On the other hand, the ultimate strength was higher in the 0% sand, which regained partial elastic behavior. The ultimate strength decreased with the sand content: 30% and 40% were close, while 50% decreased by 20%. Thus the stiffer material possessed lower ultimate strength. While the increase in temperature resulted in decrease in stiffness as anticipated, due to homogenous distribution and thus typical plastic behavior (could be compared to 0% sand), a non-typical relation was found as the increase in temperature resulted in decrease in ultimate strength, while more homogeneity should have displayed more ultimate strength. This behavior could only be explained by increase in internal pressure due to disappearance of voids which might distribute the loads, so the melting down of the voids increased the internal pressure and thus escalated the rate of failure. Overall, the plastic reject material without sand is less stiff; addition of sand increases stiffness but only to an extent that does not affect the ultimate strength.

Abrasion

The procedure of this test was conducted as previously described: the samples were subjected to abrasion simulation using sea water media and chrome balls. The samples were loaded for 24 hours, after which the machine was stopped, and the samples were weighed in air and in water. The machine was re-operated after one hour. The cycle was repeated four times. Table 4.18 shows the results and the volume difference during each cycle as per equations (6) and (7).

Table 4.18. Abrasion test results (El-Haggar and Ibrahim, 2009)

	Sample	Weight in air (gm)	Weight in water (gm)	Volume vt (m³)	Volume lost (m³)
Cycle 0	1	490.94	30.3	0.000462	0
	2	523.35	80.6	0.0004441	0
Cycle 1	1	492.94	38.3	0.000456	6.02E-06
	2	525.93	70.6	0.0004567	-1.26E-05
Cycle 2	1	492.97	30.9	0.0004635	-1.43E-06
	2	525.71	86.6	0.0004404	3.65E-06
Cycle 3	1	490.96	33.9	0.0004584	3.59E-06
	2	523.43	83.4	0.0004414	2.73E-06
Cycle 4	1	493.03	33.2	0.0004612	8.12E-07
	2	525.45	83.6	0.0004432	9.03E-07

The volume lost at the end of the cycles is fairly small: sample 1 lost about 0.176% of its original volume, while sample 2 lost about 0.2% of its original volume. The two samples did not lose significant volumes. The weighing of samples in water could be a source of error due to frequent flipping of digits in the digital meter.

Wet–dry test
The test was conducted for the samples as per the ASTM standards: the samples were immersed in water for 12 hours and then dried in an oven at 60 to 70°C for six hours. The cycle was repeated five times. This test determines the percent loss as per equation (4). The results are given in table 4.19.

Table 4.19. Average percentage loss distribution (El-Haggar and Ibrahim, 2009)

Mix	% loss
Mix 1	0.0198
Mix 2	0.0114
Mix 3	-0.0008
Mix 4	0.0298
Mix 5	0.0176

The percent loss is very small. However, it decreased with the increase of sand content. This could be due to the fact that the increased filler material prevents samples from being dissipated, leading to decrease in percent loss. When the mixing temperature is increased, some polyethylene bonds might have broken, resulting in the increase in percent loss, especially at the 160°C level.

Soundness test

The solution was supposed to cover the sample by 1/2 inch, but the samples floated. The first cycle was done by immersion of samples for 16 hours in the solution of magnesium sulfate, followed by drying for 4 hours at 60°C; however, the standards called for drying the samples at 110°C for two to four hours. Therefore, in the second cycle, the samples were subjected to 110°C for two hours. The result was that the samples melted.

The test was repeated with new samples produced from plastic rejects mixed with sand. The samples were subjected to 16 hours' immersion in the sulfate solution and drying at 50–60°C for two hours; the temperature was selected in reference to the wet–dry test and within operational temperature of the drying furnace. In accordance with the ASTM standards, the percent loss was calculated; the data are presented in table 4.20.

As predicted, the sulfate solution had no impact on the rejects, and no loss was determined; in fact, the samples gained some weight due to sulfate precipitation.

Table 4.20. Soundness test data (El-Haggar and Ibrahim, 2009)

Sample	Weight prior (gm)	Weight dried (gm)	% loss
1	1273.57	1274.51	-0.074
2	1262.88	1264.04	-0.092

Leachate

The sampled water was analyzed to evaluate the effect of leachate from plastic reject products on water quality. Table 4.21 presents the results. The parameters were normalized per mass of sample (kg). The sample in sea water weighed 1342g and the sample in fresh water 1624g.

The results were compared to EPA surface water discharge criteria. The TDS in the sea water was nearly doubled due to evaporation of sea water (more than 50–70% evaporated); thus salt concentrated, leading to the increase in TDS of the sea water sample.

The COD was within the EPA limits of 200 mg/l. The heavy metals are within Egyptian limits for discharge on surface water; however, Pb

Table 4.21. Water analysis result (El-Haggar and Ibrahim, 2009)

Parameter	Original sea water	Sea water after experiment	Fresh water after experiment	Normalized sea water (per kg)	Normalized fresh water (per kg)	EPA
pH	7.99	7.83	8.05			6–9
TSS (mg/l)	3	19	17			
TDS (g/l)	69.6	120.9	1.252	90.03	0.77	
Nitrate (mg/l)	0.44	3.96	0.88	2.95	0.542	10
Nitrite (mg/l)	0.01	0.053	0.083			
COD			117		72.03	200
Cadmium (µg/l)	2.7	0.8	1.5	0.596	0.924	40
Chromium (µg/l)	34.5	34.9	89.4	25.99	55.04	570
Lead (µg/l)	50.2	234.6	36.5	174.71	22.47	210

in sea water was slightly higher than the EPA limits. It is also important to note the inconsistency of contamination of rejects from various sources in landfills, so it would be advisable to wash the rejects prior to processing so that the washing water could then be treated, thus ensuring minimal toxicity leaching to the surface water. An increase in weight of 0.2% and 0.3% was noticed after sample immersion for 28 days in fresh and sea water respectively.

Other Applications
After the development of bricks, interlocks, manhole covers, and breakwaters, the idea of other products emerged, such as table toppings. A mold was manufactured with the required dimensions to produce a table top. Polyurethane coating is one of several coatings that could be used to improve the appearance and product quality.

Grani top layers, a mixture of marble and granite (as will be discussed in detail in chapter 8), uses granules from industrial wastes of rocks and the marble industry to improve the quality and the appearance of the top layer. Crushing of rocks is done by ball crushers (ball mill), followed by a process of sieving mixed with organic polymer and poured into another mold with 1cm space all around.

Table toppings proved to be very competitive with many coatings. Each coating was used in a different application. The grani layer proved to be the optimum coating, due to its durability and appearance. This led us to produce plates with thickness ranging from 10 to 30mm to be used for fences, garden furniture, and so on.

Another application for plastic rejects is wheels in slow-moving cars. The cost of these wheels is 10% of ordinary wheels. They have the disadvantage of dipping when carrying heavy loads on unpaved roads, but are practical for use on paved roads.

Speed bumps are another application of plastic rejects, used to control car speed in residential areas, schools, university campuses, and hospitals. Traditionally, speed bumps are made from cast iron, rubber, or asphalt mix. The new material made out of sand–plastic mix can be used to replace the cast iron or the rubber. It provides an excellent alternative from the technical, economic, and environmental points of view.

Questions

1. Estimate the quantity of plastic wastes in your country and state the percentage of the total amount of plastic waste that consists of plastic rejects.
2. Explain the current practice for disposing of plastic rejects in your country and how much it costs per one ton of plastic rejects.
3. Discuss the composition of plastic rejects.
4. Why can plastic rejects not be recycled or utilized?
5. Compare traditional manhole covers made out of cast iron and manhole covers made out of plastic rejects, from the technical, social, economic, and environmental points of view.
6. What type of products do you recommend to be produced from plastic rejects according to demand and need in your community? Support your answer with a market study.
7. Propose a semi-automated plastic rejects recycling technology. Create a feasibility study, a technical study, and an environmental study to develop a plastic rejects recycling plant.

References

Abou Khatwa, M.K., H. Salem, and S.M. El-Haggar. 2005. "Building Material from Waste." *Canadian Metallurgical Quarterly* 44, no. 3.

ASTM C88-99a. 1999. *Standard Test Method for Soundness of Aggregates by Use of Sodium Sulfate or Magnesium Sulfate*. Philadelphia: ASTM Standards Source.

ASTM C136-96a. 1998. *Test Method for Sieve Analysis of Fine and Coarse Aggregates*. Philadelphia: ASTM Standards Source.

ASTM C1138-97. 1997. *Standard Test Method for Abrasion Resistance of Concrete (Underwater Method)*. Philadelphia: ASTM Standards Source.

ASTM D543-95. 1998. *Standard Practices for Evaluating the Resistance of Plastics to Chemical Reagents*. Philadelphia: ASTM Standards Source.

ASTM D570-95. 1998. *Standard Test Method for Water Absorption of Plastics*. Philadelphia: ASTM Standards Source.

ASTM D570-98. 1998. *Standard Test Method for Water Absorption of Plastics*. Philadelphia: ASTM Standards Source.

ASTM D695-02. 2002. *Standard Test Method for Compressive Properties of Rigid Plastics*. Philadelphia: ASTM Standards Source.

ASTM D695M-91. 1998. *Standard Test Method for Compressive Properties of Rigid Plastics (Metric)*. Philadelphia: ASTM Standards Source.

ASTM D790M-93. 1998. *Standard Test Methods for Flexural Properties of Unreinforced and Reinforced Plastics and Electrical Insulating Materials (Metric)*. Philadelphia: ASTM Standards Source.

ASTM D792-00. 2000. *Standard Test Method for Density and Specific Gravity (Relative Density) of Plastics by Displacement*. Philadelphia: ASTM Standards Source.

ASTM D792-91. 1998. *Standard Test Methods for Density and Specific Gravity (Relative Density) of Plastics by Displacement*. Philadelphia: ASTM Standards Source.

ASTM D1525-96. 1998. *Standard Test Method for Vicat Softening Temperature of Plastics*. Philadelphia: ASTM Standards Source.

ASTM D2240-97. 1998. *Standard Test Method for Rubber Property: Durometer Hardness*. Philadelphia: ASTM Standards Source.

ASTM D4992-94. 2001. *Standard Test Method for Evaluation of Rock to be Used for Erosion Control*. Philadelphia: ASTM Standards Source.

ASTM D5240-04. 2004. *Standard Test Method for Testing Rock Slabs to Evaluate Soundness of Riprap by Use of Sodium Sulfate or Magnesium Sulfate*. Philadelphia: ASTM Standards Source.

ASTM D5313-04. 2004. *Standard Test Method for Evaluation of Durability of Rock for Erosion Control under Wetting and Drying Conditions*. Philadelphia: ASTM Standards Source.

ASTM Standards Source. 1998a. *Standard Specification for Industrial Floor Brick*. Philadelphia: ASTM Standards Source.

ASTM Standards Source. 1998b. *Standard Specification for Pedestrian and Light Traffic Paving Brick*. Philadelphia: ASTM Standards Source.

ASTM Standards Source. 1998c. *Standard Specification for Solid Concrete Interlocking Paving Units*. Philadelphia: ASTM Standards Source.

Bakker, P., A. van den Berge, R. Hakenberg, M. Clabbers, M. Muttray, B. Reedijk, and I. Rovers. 2003. "Development of Concrete Breakwater Armour Units." First Coastal, Estuary and Offshore Engineering Specialty Conference of the Canadian Society of Civil Engineering, June.

BS EN 124. 1994. *European Standards for Gully Tops and Manhole Covers for Vehicular and Pedestrian Users*. London: European Standard Source.

Burcharth, H.F., and S.A. Hughes. 2003. "Coastal Engineering Manual: Types and Functions of Coastal Structures." EM 1110-2-1100, part 5, ch. 2. US Army Corps of Engineers.

Clark-Drain Drainage Company. 2011. "Access Covers and Manhole Covers." www.clark-drain.com/productrange-castings.php?l=en, 6-7-2011

Cordon, W.A. 1979. *Properties, Evaluation, and Control of Engineering Materials*. New York: McGraw-Hill.

CWC (Clean Washington Center). 1996. "Beneficial Reuse of Spent Foundry Sand." Technology Brief, August. www.cwc.org/industry/ibp951fs.pdf

Enotes. "Manhole Cover." 2009. www.enotes.com/how-products-encyclopedia/manhole-cover

EPA (Environmental Protection Agency). 2005. *Water Quality Criteria*. www.epa.org/waterscience (May 2006).

ES (Egyptian Standard for Cement Tiles). 1974. *Book of Egyptian Standards*. Cairo: Egyptian Organization for Standardization and Quality.

El-Haggar, S.M. 2007. *Sustainable Industrial Design and Waste Management: Cradle to Cradle for Sustainable Development*. New York: Elsevier Academic Press.

El-Haggar, S.M., and L. El-Hatow. 2009a. "Standard Manhole Covers from Thermoplastic Rejects." Ain Shams University Third International Conference on Environmental Engineering (ASCEE-3), 14–16 April.

El-Haggar, S.M., and L. El-Hatow. 2009b. "Thermoplastics Reinforced with Foundry Sand Waste." *ASCE, Practice Periodical of Hazardous, Toxic, and Radioactive Waste Management Journal* 13, no. 4 (October).

El-Haggar, S.M., and Y.M. Ibrahim. 2009. "Utilization of Rejected Waste in Producing Shoreline Erosion Protection Structures (Breakwaters)." Ain Shams University, Third International Conference on Environmental Engineering (ASCEE-3), 14–16 April.

El-Hatow, L.M. 2006. "Utilization of Thermoplastic Rejects with Different Reinforcing Elements in the Production of Manhole Covers." M.Sc. thesis, American University in Cairo.

Holliday, L., ed. 1966. *Composite Materials*. Amsterdam: Elsevier.

Klieger, P., and J.F. Lamond, eds. 1994. *Significance of Tests and Properties of Concrete and Concrete Making Materials*. Philadelphia: ASTM.

Kourany, Y., and S.M. El-Haggar. 2001. "Using Slag in Manufacturing Masonry Bricks and Paving Units." *TMS Journal* 9: 97–106.

Morningrose. N.d. "GRP Manhole Cover." Made in China.com. morningrose.en.made-in-china.com/product/oeXJMCqdkQkl/China-GRP-Manhole-Cover-006-.html

Muir, H. 2012. "Rising Ripoffs." *The Guardian*, 31 July. www.guardian.co.uk/uk/2004/oct/25/ukcrime.prisonsandprobation

Nielsen, L.E., and R.F. Landel. 1994. *Mechanical Properties of Polymers and Composites*. 2nd ed. New York: Marcel Dekker.

Reno, J. 2012. "Rising Ripoffs." Newsweek.com, 4 August. www.newsweek.com/id/137822

Shackel, B. 1990. *Design and Construction of Interlocking Concrete Block Pavements*. New York: Elsevier Applied Science.

USNA (United States Naval Academy). 2006. "Breakwaters." www.usna.edu

Whisprwave® Wave Dispersion Technologies Inc. 2005. "Floating Breakwaters Provide Erosion Control Protection and Wave Attenuation." www.whisprwave.com

Innovation in the Plastic Industry 2: Upcycle of Plastic Waste: Natural Fiber–Reinforced Plastics

Irene Fahim and Salah M. El-Haggar

Introduction

Plastics are organic materials that have the ability to flow into a desired shape under the application of heat and pressure, and to retain that shape when the heat and pressure are withdrawn. Plastics were first invented in 1862, but the material was called Parkesine. It was an organic material derived from cellulose that once heated could be molded, and retained its shape when cooled. Yet plastics have only been widely used in the last fifty years. They are light, durable, moldable, hygienic, and economical. Because of these features, they have been utilized in product packaging, car manufacturing, pillows, mattresses, computers, wires, electric cords, and agricultural and housing products. Plastics are composed of long chain molecules called polymers, derived from naturally occurring substances such as coal, natural gas, and oil. These polymers can then be made into granules, powders, and liquids, becoming the raw materials for plastic products (Bellis, 1997).

Plastic production worldwide now exceeds 200 million tons a year. The plastics industry involves manufacturing polymer materials including natural, synthetic, and organic compounds, building construction, bulk material handling, packaging, and transportation, as shown in figure 5.1. It continues to have a huge growth potential. Plastics and polymer consumption will have an average growth rate of 5% per year, reaching 297 million tons by 2015, as shown in table 5.1 (Nkwachukwu et al., 2013).

Figure 5.1. Some of the types of plastic (cipet.gov, 2010)

The plastic industry can be categorized into upstream and down-stream manufacturing. The upstream side involves the manufacture of polymers by large manufacturing plants utilizing state-of-the-art technology. The downstream side is the conversion of polymers into plastic products. These companies are usually of micro, small, and medium scale, accounting for only 25% of polymer consumption. The industry also consumes recycled plastic, which constitutes about 30% of total consumption (El-Haggar, 2007). Even though plastics have a variety of benefits, they are in fact detrimental to the environment; plastic production consumes a large amount of energy and materials, primarily fossil fuel that when combusted emits hazardous toxins into the air. It is estimated that 4% of the world's annual oil production is used as feedstock for plastics production and an additional 3.4% during manufacture. Plastic production also involves the use of potentially harmful chemicals, which are added as stabilizers or colorants. Many of these have not undergone environmental risk assessment and their impact on human health and the environment is currently uncertain. An example of this is phthalates, which are used in the manufacture of polyvinylchloride (PVC). PVC has in the past been used in toys for young children and there has been concern that phthalates may be released when these toys come into contact with saliva, if the toy is placed in the child's mouth or chewed by the child. Risk assessments of the effects on the environment are currently being carried out. Other environmental impacts of plastics include the extensive amount of water that is needed in manufac-

Table 5.1. Global per capita consumption of plastics in 2012 (Nkwachukwu et al., 2013)

Country	kg/capita
World average	26
North America	90
Western Europe	65
Eastern Europe	10
China	12
India	5
Southeast Asia	10
Latin America	18

turing. The numerous plastic bags that are dispersed as litter in urban areas have also become a serious concern. Due to the magnitude of the problems associated with plastic, the world is now turning toward the use of recycled plastics. There are numerous benefits to the recycling of plastics: reducing water usage by 90%, reducing CO_2 emissions by 60%, and reducing energy consumption by two-thirds (El-Haggar, 2007).

Types of Plastic

Plastics are classified according to the chemical structure of the polymer's backbone and side chains. Acrylics, polyesters, silicones, polyurethanes, and halogenated plastics are some of the main groups. Plastics can also be classified by the chemical process used in their synthesis, including condensation, polyaddition, and cross-linking. The two basic types of plastics are thermoplastics and thermoset polymers. Thermoplastics are those that do not undergo chemical change in their composition when heated and can be molded again and again, including polyethylene, polypropylene, polystyrene, and PVC, as shown in table 5.2. Their chains are composed of several thousand repeating molecular units derived from monomers, also known as repeat units. Thermosets, on the other hand, can melt and take shape only once; after they solidify, they stay solid. Therefore, the chemical reaction of the thermosetting process is irreversible. The vulcanization of rubber is an example of a thermosetting process: before heating with sulfur, the polyisoprene is a tacky, slightly runny material, but after vulcanization the product is rigid and non-tacky. Plastics can also be classified based on qualities relevant to manufacturing or product design. Besides thermoplastic and thermoset, these include elastomeric, structural, biodegradable, and electrically conductive. Yet another classification system is by physical properties, such as density, tensile strength, glass transition temperature, and resistance to various chemical products (USEPA, 2006).

Biodegradable plastics break down upon exposure to natural or environmental factors such as sunlight (e.g., ultraviolet radiation), water or dampness, bacteria, enzymes, or wind abrasion, as well as rodent or

Table 5.2. Types of polymers and their application (USEPA, 2006)

Polymer types	Examples of applications	Symbol
Polyethylene terephthalate	Fizzy drink and water bottles Salad trays	1 PETE
High-density polyethylene	Milk bottles, bleach, cleaners, and most shampoo bottles	2 HDPE
Polyvinyl chloride	Pipes, fittings, window and door frames (rigid PVC) Thermal insulation (PVC foam) and automotive parts	3 V
Low-density polyethylene	Carrier bags, bin liners, and packaging films	4 LDPE
Polypropylene	Margarine tubs, microwave-able meal trays: also produced as fibers and filaments for carpets, wall coverings	5 PP
Polystyrene	Yogurt pots, foam hamburger boxes, and egg cartons	6 PS

insect attack. Degradation can occur for plastics exposed at the surface, or if certain conditions exist in landfill or composting systems. Degradation can be enhanced by adding starch powder to plastic, as filler. However, this does not lead to complete degradation. Expensive biodegradable plastic, such as Biopol, has been synthesized from genetically engineered bacteria. The German chemical company BASF has fabricated another biodegradable material named Ecoflex, a fully biodegradable polyester for food packaging applications. A replacement for this expensive biodegradable material is natural fiber–reinforced plastic (USEPA, 2006).

Annual average growth rates (AAGR%) for all polymers, plus PP (polypropylene), PE (polyethylene), PVC, and PS (polystyrene), are depicted in figure 5.2. Demand for the various grades of polyethylene is predicted to grow an average 4.4% annually through 2020 (fig. 5.3). LLDPE is expected to experience the fastest growth, with an average annual growth rate of 6.2 percent, followed by LDPE, which is expected to grow only 1.8% during the same period, as shown in figure 5.3 (Nkwachukwu et al., 2013).

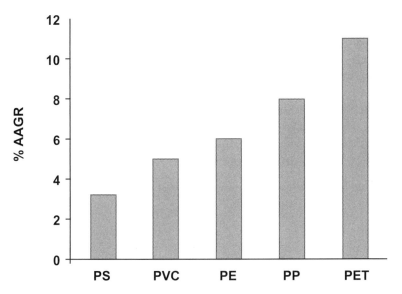

Figure 5.2. Annual average growth rate of polymers (Nkwachukwu et al., 2013)

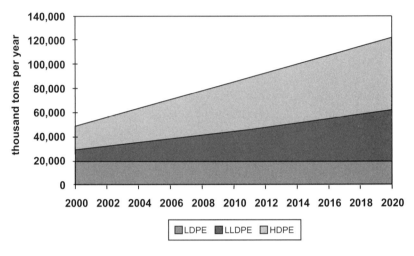

Figure 5.3. Global polyethylene demand growth (1000 tons/year) (Nkwachukwu et al., 2013)

Plastic Waste

The market potential for the usage of plastic waste is huge, because of the high amounts of waste generated: 227 million tons of plastic waste were generated in 2012 (Nkwachukwu et al., 2013). This includes municipal solid waste (60%), residential (11%), commercial, industrial, and health care wastes (29%). The majority of plastic waste generated is disposed. However, plastic consumption is continuously growing, due to its short life cycle—about 40% of it has a life span of less than one month. For this reason, plastics should be recovered, rather than disposed or incinerated, so as to minimize landfill content and the potential negative environmental impacts associated with its disposal, including air and land pollution. Besides, not closing the cradle-to-cradle loop and depleting natural resources both go against the concept of sustainable development (PEMRG, 2011).

Even though technological advances for recycling plastics have soared, there remains the unanswered question of cost. Do the economics of the process balance out and pay off in the end? According to an article in *Geographical* titled "The Cost of Recycling," the answer is no (PEMRG, 2011). The article states that "more money is spent recycling plastic products than is reclaimed in the process. The resulting extracted plastics can be used for low-grade products with little market value." Plastics degrade during recycling and can only undergo the process once (Palmer, Sigman, and Walls, 2007). However, a study entitled "The Cost of Reducing Municipal Solid Waste" done by Resources for the Future Foundation shows that plastic recycling does in fact pay back when it is done in large quantities. If it is done for a whole community or district, the paybacks may be more obvious. Yet when the quantities are much lower, the cost becomes an obstacle due to the expenses of energy, transportation, sorting, and labor. The importance of recycling plastics on the larger scale cannot be overstated, as its benefits are numerous (Pickering, 2008).

Natural Fibers

Natural fibers are composites of hollow cellulose fibrils held together by a lignin and hemicelluloses. Each fibril consists of a thin primary wall around a thick secondary wall. The secondary wall is made up of three layers, including a thick middle layer. This middle layer determines the mechanical properties of the fiber. It consists of a series of helically wound cellular micro fibrils made of long-chain cellulose molecules (Bachtiar, Sapuan, and Hamdan, 2008). Table 5.3 shows the properties of several natural fibers.

Table 5.3. Mechanical properties of natural fibers for composite applications (Bachtiar, Sapuan, and Hamdan, 2008)

	Tensile strength (MPa)	Elongation at break (%)	Young modulus (GPa)
Flax	300–1500	1.3–10	24–80
Jute	200–800	1.16–8	10–55
Sisal	80–840	2–25	9–38
Coir	106–175	14.21–49	4–6
Oil palm	130–248	9.7–14	3.58
Hemp	310–900	1.6–6	30–70
Wool	120–174	25–35	2.3–3.4
Cotton	264–800	3–8	5–12.6
Rice straw	69–100	9.4–12	27–30

Treatment of natural fibers

Natural fibers are chemically or thermally treated to be used in different applications as a reinforcing medium for polymers or other materials. The main objective of the treatment is to remove impurities. The chemical treatment uses acids, alkali, and carbonization to enhance the mechanical properties of the fibers and remove the impurities. The enhancement in mechanical properties, including flexural and tensile strength in alkali-treated fibers, is attributed to improved wetting of alkali treated with the polymer matrix. The fibers that can be treated with alkali solutions are jute, palm, wood, and paper. The following experiments were conducted to test various treatments of fibers.

An aqueous solution of 5% NaOH was applied for two hours at 150°C, and then washed with distilled water until the NaOH was eliminated. The fibers were dried at 60°C for 24 hours. The increase in the chemical concentration of the alkali damaged the fiber structure. Soaking for longer gave better results due to improved crystallinity of fibers and removed hemicellulose and lignin (Maya and Sabu, 2004). Another alkali treatment for sisal fibers was to apply an NaOH solution of between 4 and 10% for one hour at room temperature. The diameter and weight of fibers decreased due to removal of lignin.

Adding sodium hydroxide and starch ethylene vinyl alcohol to 10g of coir fibers improved the thermal stability of fibers, compatibility, and interfacial bond strength. Tensile strength improved by 53% due to lack of impurities. Treatment in water improved the values of Young's modulus by 75%. The treatment also improved thermal degradation of fibers. Improvements can be attributed to the removal of hydrolyzed substances, which decompose earlier than cellulose lignin, leading to higher thermal stability at the second stage of degradation. Crystallization of fiber increases due to rupture of linkage lignin and alkali (Haque and Hasan, 2009). Another treatment is adding 1% concentration of NaOH for three hours at room temperature to bagasse, then washing, drying at 100°C for 72 hours, and cleaning. The phenomenon of split fibers breaks fiber bundles into smaller ones by dissolution of hemicelluloses. This increases the effective surface area available for contact with matrix, thus improving interfacial bond. Pullout effect decreased in treated fibers, and the composite system displayed reduced water uptake (Brígida, Calado, and Coelho, 2010).

Coconut fiber (2–5g) was chemically treated using three methods: sodium hypochlorite (NaOCl), sodium hypochlorite/sodium hydroxide

(NaOCl/NaOH), or hydrogen peroxide (H_2O_2). The effect of these treatments on the structure, composition, and properties of fibers was studied. The advantage of treatment with H_2O_2 was that it was most efficient in the removal of waxy and fatty acid residues; however, it does not modify the surface chemical composition. Treatment with NaOCl/NaOH reduced the hemicellulose content in the fibers, which showed a greater exposure of cellulose and a reduction in thermal stability (Baek and Kwon, 2007).

Four concentrations of sulfuric acid (H_2SO_4) were used (1–4%) on 5g of wheat straw at three different temperatures (98, 121, and 134°C). The predominant effect was to solubilize the hemicellulose fraction of wheat straw. This was achieved by using the highest temperature. The solubility reached 67% but there was no significant improvement in the thermal stability and interfacial bond. However, there was an improvement in tensile stress of the fiber-reinforced polymer by 40–50% with the 1% concentration of H_2SO_4. This result promotes a cleaner production technique. Using 1% instead of 4% leads to reduction of chemicals at the source, thus minimizing cost. Another treatment of natural fibers was done using a sulfuric acid concentration of 1.5% for microbial production of Xylitol (Castro, 1993).

Certain physical and chemical treatments can be applied to natural fibers to overcome their natural deficiencies. Some natural fibers have poor compatibility with some polymeric matrices when wet, displaying high moisture absorption and poor adhesion. This means that the full capabilities of the composite cannot be activated and it is prone to degradation by environmental elements. This weakens the composite and reduces its life span. There are two physical treatments that can be applied: cold plasma and corona treatment. The latter is a technique for activating surface oxidation by changing the surface energy of the fiber. The chemical treatment can be done using maleic anhydride, organosilanes, isocyanates, sodium hydroxide, permanganate, and peroxide. These coupling agents eliminate weak boundary layers, produce a tough, flexible layer, develop a highly cross-linked interphase region, improve the wetting between polymers, and form covalent bonds with both materials. The tensile strength of the maleic anhydride-treated fiber composites was found to be higher than untreated fiber. The jute–epoxy composite, when treated with ilane, proved to have better interfacial bonding and hence better mechanical properties, especially adhesion. Another example of chemical treatment is adding wax. Wax enhances natural fiber-reinforced plastics by overcoming processing problems caused by cross–linking and decomposition. The wax has low velocity to enable impregnation. The mechanical properties of composites increased with these treatments, but any increase in treatment results in higher costs (Ngoc, 2006).

Thermal treatment of natural fiber through carbonization is another method, in which heating without air occurs. The pore structure and adsorption properties of carbonized fibers are utilized in several applications, as they result in improved mechanical properties. Carbonization starts with thermal treatment of raw fibers in an inert atmosphere, followed by an activation step with CO_2 or steam at the same temperature. In chemical activation, the raw fibers are impregnated in a solution of phosphoric acid and heated at 900°C in an inert atmosphere. Chemical treatment leads to high porosity, which enables a high adsorption capacity for micro pollutants. Moreover, it produces numerous acidic surface groups involved in the adsorption mechanism of dyes and metal ions. The phosphoric acid increases the carbon yield by promoting the bond cleavage in the biopolymers and dehydration at low temperatures, followed by extensive cross-linking that binds volatile matter into the carbon product (Seong and Kim, 2008).

Natural Fiber–reinforced Plastics (NFRP)

Fiber-reinforced plastics (FRP) are useful in a number of industrial products, including underground pipes, water supply pipes, rag pipes, and shield pipes for communication cables. Traditionally, damaged underground pipes are repaired by excavation and replacement of damaged sections. A solution that reduces the time and cost of excavation is vacuum-assisted resin transfer molding. The drawbacks of this technique include interior defects and long resin injection time; however, the rehabilitation process is more reliable. The material used for the new pipes is natural fiber–reinforced composite; the natural fiber used is jute, which is moderately priced and highly resistant to corrosion. It has the further advantage of avoiding contamination of water in the pipes, unlike glass fibers. Jute has a low tensile strength, so placing glass fiber along the outer surface of the pipes and jute at the inner surface is recommended to increase the impregnation speed and strength of the pipes (fig. 5.4). Experience has shown that this combination is not only environmentally friendly, but reduces the operation time by half and satisfies the ASTM standard, as the glass provides a smooth passage for

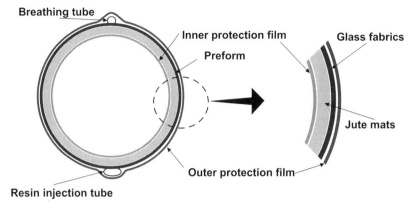

Figure 5.4. Reinforcement with glass fabric and jute mat (American Concrete Institute, 1996)

the resin while the jute provides permeability under applied pressure. The cost of jute is $0.46/m^2$, while glass costs $1/m^2$ (American Concrete Institute, 1996).

Another application for FRP is in highway construction. Highway infrastructure has been deteriorating for many years, due to a number of factors including harsh environmental conditions, heavy loads, insufficient maintenance, high traffic volumes, and tight construction budgets. Furthermore, limitations in roadway construction areas have placed constraints in terms of the need for rapid construction, durable structural components, and lightweight, easily constructed facilities. One possible solution to these problems is fiber-reinforced polymer (FRP) composite materials, which have been used for some time in the aerospace and military communities. FRP's strengths overcome the limitations of several traditional materials. Research has been conducted during the past decade, in the United States and other countries, on the use of FRP composite materials for highway infrastructure applications (Van Den Einde, Karbhari, and Seible, 2007).

FRP composites are useful in civil engineering applications due to their high strength-to-weight and stiffness-to-weight ratios, corrosion resistance, light weight, and potentially high durability. Important applications include the renewal of constructed-facilities infrastructure such as buildings, bridges, and pipelines, and the rehabilitation of concrete structures—mainly because of their tailored performance characteristics, ease of application, and low life-cycle costs. The success of these materials in structural rehabilitation applications has led to the development of new lightweight structural concepts utilizing all-FRP systems or new FRP/concrete composite systems (Baker et al., 1991).

FRP doors can be produced in different sizes, have easy workability, and are durable, impact-resistant, water-repellent, chemical-resistant, and stainproof. FRP roof rainwater gutters are fitted on the roofs of industrial buildings for water drainage. FRP gratings are used in chemical, beverage, aircraft, and food applications; they have non-skid surfaces which are non-corrosive, non-magnetic, non-conducive, lighter than aluminum, and 50% stronger than hot rolled steel. They also have low thermal expansion compared to aluminum and steel, and are easy to install and maintain (Baker et al., 1991).

Testing of processing technique for NFRP

Rice straw fibers were cut to three lengths (2mm, 4mm, and 6mm) through a shredder. The fibers were treated with 1% and 5% phosphoric acid, sulfuric acid, and sodium hydroxide. The scale used to weigh the fibers was calibrated using the linear regression method based on standard weights. The fiber was soaked for two hours at 150°C in acid and alkali solutions, then washed with distilled water. The treated fiber was dried in a furnace at 60°C and the moisture was totally eliminated. The fibers were weighed every hour. After eight hours, the weight did not change, guaranteeing that the fibers were dry. This procedure was repeated three times to verify the results. The sample preparation procedure shown in figure 5.5 started by shredding the fibers. The fibers and polymer were mixed using a mixer, then fed into a single screw extruder as shown in figure 5.6. The mix was placed in the hopper of the extrusion machine and temperatures were set at 120°C for the first heater and 150°C for the second heater in the extruder. The first heater heated the mix to the desired temperature and properly melted it, while the second heater ensured the flow of melted plastic with the additives before extrusion of the paste from the die. The heaters were insulated and covered with glass wool to minimize heat loss. The cooling section was located at the beginning of the extruder with the water inlet and outlet to guarantee there was no melting in the mixing or feeding chamber. The mix was fed gradually so as not to clog up the rotating screw at the input point of the machine. The process produced 500g of extrudates within seventeen minutes. The paste was pressed in a hydraulic press. A custom-made steel die with specific

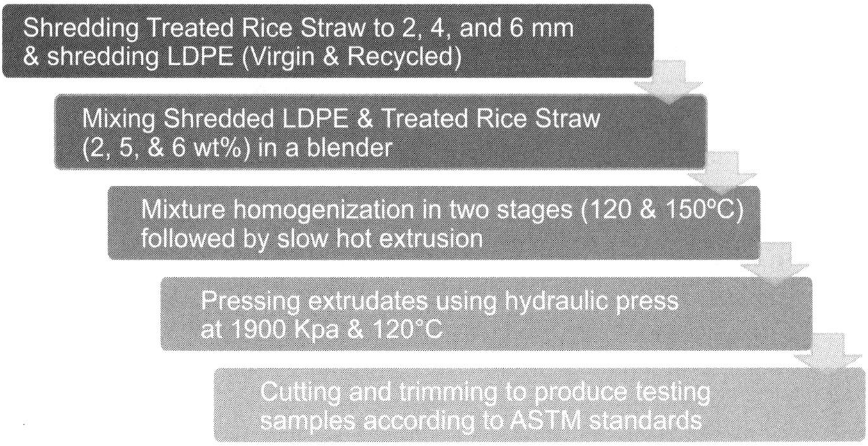

Shredding Treated Rice Straw to 2, 4, and 6 mm
& shredding LDPE (Virgin & Recycled)

Mixing Shredded LDPE & Treated Rice Straw
(2, 5, & 6 wt%) in a blender

Mixture homogenization in two stages (120 & 150°C)
followed by slow hot extrusion

Pressing extrudates using hydraulic press
at 1900 Kpa & 120°C

Cutting and trimming to produce testing
samples according to ASTM standards

Figure 5.5. Sample preparation procedure (El-Haggar, 2007)

Figure 5.6. The extrusion machine (El-Haggar, 2007)

dimensions was used to satisfy the requirements of testing. Trimming
and cutting processes were done to prepare the product for testing in
accordance with the requirements of the testing standards.

Characterization techniques

There are several tests available for fiber-reinforced polymers. Flexural and tensile stress tests were chosen, as they are the most extreme stresses the product will sustain under severe usage conditions. Tensile stress is "the maximum stress that a material can withstand while being stretched or pulled before necking, when the specimen's cross-section starts to significantly contract." The ultimate tensile stress is the highest point of the stress–strain curve that is produced when a tensile test is performed and the stress-versus-strain is recorded (ASTM D3039/D3039M, 2008). The machine used for this procedure was a 3300 Instron, as shown in figure 5.7. Instron is a testing machine that can conduct tensile, compression, flex, and peel and cyclic types of testing. The two edges of the machine consist of the support span, where the grips are attached to the load ram and the specimen is concentrated between the support edges. After adjusting the computer software settings, the machine is ordered to start the test and the ram moves downward, applying the load. The output of the test is a stress–strain diagram. In our experiment, tension samples were selected according to ASTM D3039/D3039M. Mechanical testing of these chemically treated natural fibers included tensile and flexural tests. The standard specimens were tested at a rate of 2mm/min–5mm/min. The flexural stress, failure strain, and tensile stress were calculated from a stress–strain curve. Sample standard dimensions according to ASTM were 40–100mm length, 10–20mm width, and 2–5mm thickness (ASTM D3039/D3039M, 2008). Grip was calculated at 20mm and crosshead speed at 5mm/min.

Figure 5.7. Instron Universal Testing Machine with tension grip

Figure 5.8. Flexural fixture with three-point bending

The flexural test is performed with a three-point loading, in which the specimen is loaded with one nose in the middle of the specimen support span as shown in figure 5.8. The maximum axial fiber stress is positioned directly under the loading nose. The flexural stress represents the highest stress experienced within the material at its moment of rupture. It is measured in terms of stress. The stress in a rectangular sample under a load in a three-point bending can be determined by the equation

$$\sigma = 3FL/2bd^2$$

where F is the load (force) at the fracture point, L is the length of the support span, b is width, and d is thickness. Bending samples were selected according to ASTM D790-03 (2008): width = 20mm, thickness = 5mm, span = 16 x thickness, and crosshead speed = 2mm/min.

Results
The effects of fiber reinforcement on virgin LDPE and recycled LDPE using the flexural and tensile test were investigated using phosphoric acid, sulfuric acid, and sodium hydroxide.

Effect of fiber reinforcement on virgin LDPE using the flexural test
Three samples for each treatment condition were tested in accordance with ASTM D790-03 to determine the flexural stress of FRP of 2% and 5% rice straw of length 2mm treated with sodium hydroxide alkali. The increase in fiber content showed an improvement in flex-

ural stress at 5% fiber concentration. The 5% concentration of the alkali showed slightly less change in flexural stress than the 1% sample, as shown in figure 5.9. The micro-analysis shows that interfacial stress between rice straw and the LDPE matrix is the cause of good fiber reinforcement. The interface acts as a 'binder' and transfers the load between the matrix and the reinforcing fibers. Mechanical stress of the composites decreases at high fiber loading: the fiber–fiber contact dominates over the resin matrix–fiber contact, which decreases the mechanical properties.

The same procedure as for sodium hydroxide was conducted using sulfuric acid treatment for each condition. The increase in fiber concentration from 2% to 5% showed an improvement in flexural stress. The concentration of the acid showed no significant change in flexural behavior, as shown in figure 5.10.

The same procedure was again performed using phosphoric acid, with several samples repeated for each condition. The increase in fiber concentration from 2% and 5% showed an improvement in flexural stress. The different concentration of the acid showed a slight change in flexural behavior, as shown in figure 5.11.

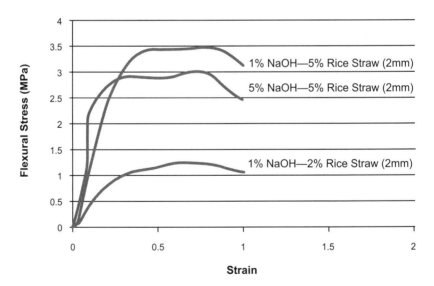

Figure 5.9. Variation of flexural stress for different fiber concentrations with 2mm length using 1% and 5% NaOH

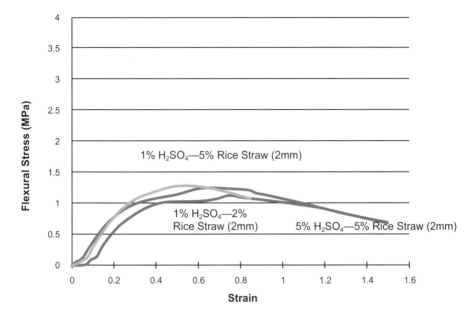

Figure 5.10. Variation of flexural stress for different fiber concentrations
with 2mm length using 1% and 5% H_2SO_4

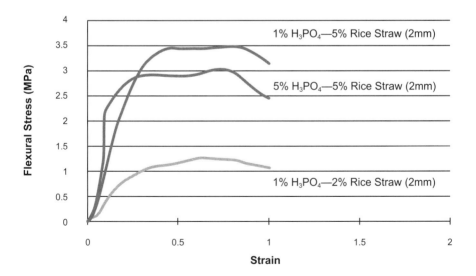

Figure 5.11. Variation of flexural stress for different fiber concentrations
with 2mm length using 1% and 5% H_3PO_4

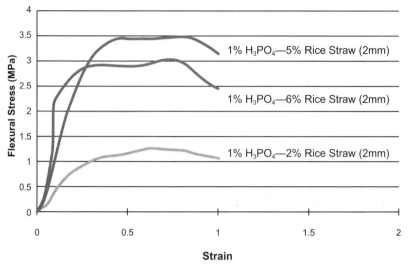

Figure 5.12. Effect of fiber length on flexural stress using 1% chemical concentration and 5% fiber concentration

The flexural stress values of NaOH-treated fiber and the H_3PO_4-treated fibers have higher flexural behavior than the H_2SO_4-treated fibers. It can be concluded that the most suitable treatment is the one involving H_3PO_4 acid with 5% fiber content and 1% fiber concentration.

The effect of 5% of treated fiber concentration using 1% chemical concentration with 2, 4, and 6mm fiber length on the reinforcement of LDPE was investigated. Flexural stresses were decreased for composites with 4mm and 6mm length compared to the 2mm fiber-reinforced composites, as shown in figure 5.12.

Effect of fiber reinforcement on virgin LDPE using the tensile test
Three samples were tested for each condition in accordance with ASTM D3039 to determine the tensile stress of FRP of 2% and 5% rice straw of length 2mm treated with sodium hydroxide alkali. The increase in fiber content showed an improvement in tensile stress at 5% fiber. Samples of 1% and 5% concentration of NaOH were also tested. The 5% concentration of the alkali showed no significant difference in tensile stress from 1%, as shown in figure 5.13.

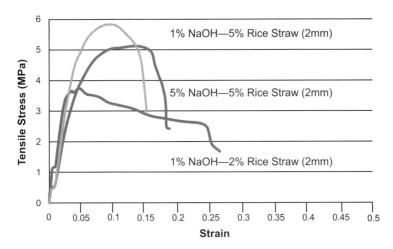

Figure 5.13. Variation of tensile stress for different fiber concentrations
with 2mm length using 1% and 5% NaOH

Three samples for each condition were tested in accordance with ASTM D3039 to determine the tensile stress of FRP of 2% and 5% rice straw of length 2mm with sulfuric acid treatment. The increase in fiber content showed an improvement in tensile stress at 5%. The 5% and 1% concentrations of the acid showed no significant change in tensile stress, as shown in figure 5.14.

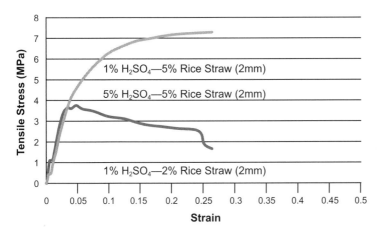

Figure 5.14. Variation of tensile stress for different fiber concentrations
with 2mm length using 1% and 5% H_2SO_4

Three samples were tested in accordance with ASTM D3039 to determine the tensile stress of FRP of 2% and 5% rice straw of length 2mm with phosphoric acid treatment. The increase in fiber content showed an improvement in tensile stress at 5% fiber. The 5% concentration of the acid showed no significant change in tensile stress from 1%, as shown in figure 5.15.

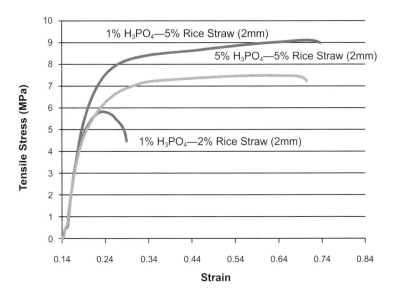

Figure 5.15. Variation of tensile stress for different fiber concentrations with 2mm length using 1% and 5% H₃PO₄

It can be concluded that phosphoric acid treatment has the highest tensile stress compared to other acids and alkali. When the effect of 5% of treated fiber concentration using 1% chemical concentration with 2, 4, and 6mm fiber length on the reinforcement of LDPE was investigated, it was found that tensile stress for composites of fiber length 4mm and 6mm decreased, compared to 2mm fiber-reinforced composites, as shown in figure 5.16.

Effect of fiber reinforcement on recycled LDPE using flexural test
The effect of chemical concentration with 2mm fiber length on the reinforcement of recycled low-density polyethylene (RLDPE) using phosphoric acid (H₃PO₄) is shown in figure 5.17. It is clear that the 1% concentration obtained higher flexural stress than the 5% concentration.

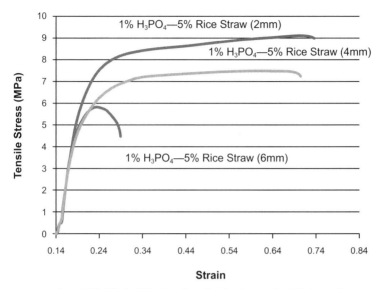

Figure 5.16. Effect of fiber length on tensile stress using 1% chemical concentration and 5% rice straw

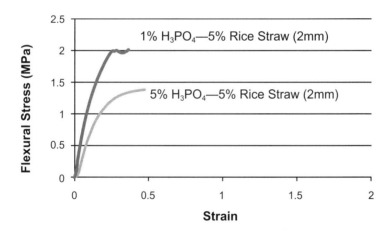

Figure 5.17. Effect of chemical concentration using H_3PO_4

Thus the former can be used in order to avoid the use of unneeded chemicals.

The effects of fiber content of 2%, 5%, and 6% are tested with 2mm length. The recycled polymer without reinforcement yielded 1.2 MPa flexural stress. The most suitable sample is 5% rice straw of 2mm length treated with 1% concentration of H_3PO_4. The increase in fiber content lowered the flexural stress, as shown in figure 5.18.

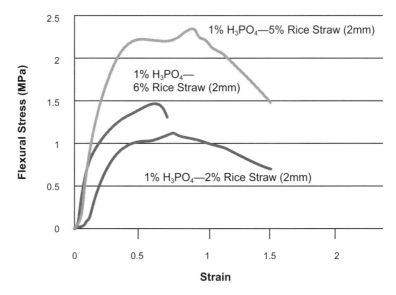

Figure 5.18. Effect of fiber content with 2mm fiber length using 1% H_3PO_4

The effect of 5% of treated fiber content using 1% chemical concentration with 2mm, 4mm, and 6mm fiber length on the reinforcement of RLDPE is noted in figure 5.19. It shows that an increase in fiber length leads to a decrease in flexural stress. At high fiber volume fractions, the material properties show greater variability due to fiber clumping. Fiber length decreases with increasing fiber content, and this reduction then shrinks fiber reinforcing efficiency due to the heightened fiber–fiber interaction and fiber–equipment wall contact. Moreover, the increase in fiber content causes the formation of agglomerates acting as a stress focus and, therefore, as a propagator of cracks, resulting in the greater tenacity of the composites.

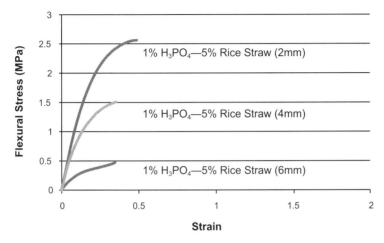

Figure 5.19. Effect of different fiber length using 5% fiber content and 1% H₃PO₄

Effect of carbonization treatment on recycled LDPE using the flexural test

Thermal treatment through carbonization is another process used on natural fibers. The effect of treated carbonized fiber was measured using 2% and 5% fiber content. The sample with treated, carbonized 5% rice straw with 2mm size has higher flexural stress compared to the samples with 4mm and 6mm size, as seen in figure 5.20. The mechanical properties of recycled LDPE reinforced with rice straw were one-fifth those of virgin composite in terms of tensile strength and half of those for flexural strength, but still higher than the recycled LDPE. This treatment contributed to the formation of fibers with higher values of surface area and micro porosity. The carbonized rice straw with the recycled LDPE resulted in six times the value of tensile strength compared to the recycled LDPE reinforced with chemically treated rice straw.

Effect of fiber reinforcement on recycled LDPE using tensile stress

The effect of chemical concentration with 2mm fiber length on the reinforcement of RLDPE using H_3PO_4 is noted in figure 5.21. The 1% concentration obtained higher flexural stress than using 5% chemical concentration. To avoid the use of unneeded chemicals, a 1% concentration should therefore be used.

The effects of fiber content of 2%, 5%, and 6% were tested with 2mm length, as shown in figure 5.22. The recycled polymer without

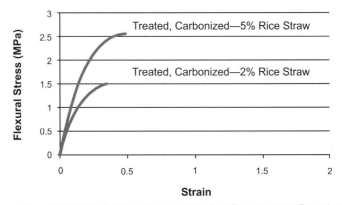

Figure 5.20. Effect of different treated, carbonized fiber content on flexural stress

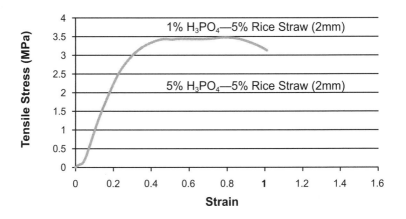

Figure 5.21. Effect of chemical concentration using H_3PO_4

reinforcement yielded 1.5 MPa tensile stresses. The increase in fiber content lowered the tensile stress.

The effect of 5% of treated fiber content using 1% chemical concentration with 2mm, 4mm, and 6mm fiber length on the reinforcement of RLDPE was investigated using the tensile test. It was found that 1% chemical concentration obtains almost the same stress as using 5% chemical concentration. Figure 5.23 shows the effect of increasing fiber length leading to a decrease in flexural stress.

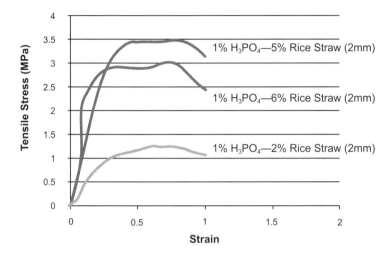

Figure 5.22. Effect of fiber content with 2mm fiber length and 1% H_3PO_4

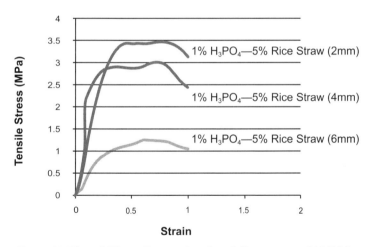

Figure 5.23. Effect of different fiber lengths using 5% fiber content and 1% H_3PO_4

Effect of carbonization treatment on recycled LDPE using the tensile test

The fibers are treated by carbonization; the effect of treated fiber is investigated using 2% and 5% fiber content. The sample with the treated, carbonized 5% rice straw displayed higher tensile stress compared to 2% fiber concentration, as can be seen in figure 5.24. There was an improvement in tensile stress from 3.5 to 9 MPa after using the

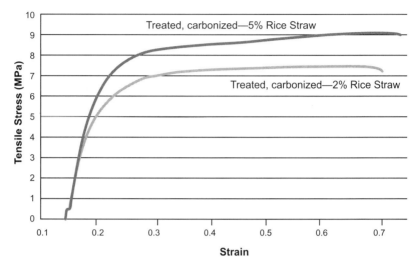

Figure 5.24. Effect of different treated, carbonized fiber content on tensile strength

carbonized rice straw with a concentration of 5%. This value showed a great improvement, reaching the value of the virgin LDPE reinforced with chemically treated fiber. The effect of treatment by carbonization in flexural stress was the same value as recycled polymer reinforced with chemically treated fiber.

Questions
1. What are the emerging trends in the global natural fiber plastic industry?
2. What are the advantages of reinforcing polymers with natural fibers? Give one application used in your country.
3. What is the effect of rising prices of petroleum-based products on the natural fiber industry?
4. Explain the theory behind the improvement of polymers when reinforced with natural fibers.
5. What is the effect of treating natural fiber before using it as reinforcement?
6. Mention several types of natural fibers and their applications in industry.
7. What are the differences between using recycled plastics and virgin plastic for FRP?

8. What are the benefits of using recycled LDPE?
9. Can polyethylene or other types of plastics be used for FRP? Explain.
10. What effect does changing the size of fiber and the method of treatment have?

References

American Concrete Institute. 1996. *State of the Art Report of Fiber Reinforced Plastic Reinforcement for Concrete Structures*. February. Detroit, MI: Woodhead Publishing.

ASTM D790. 2008. *Standard Test Method for Flexural Properties of Unreinforced and Reinforced Plastics and Electrical Insulating Materials by Four-point Bending*. Philadelphia: ASTM International.

ASTM D3039/D3039M. 2008. *Standard Test Method for Tensile Properties of Polymer Matrix Composite Materials*. Philadelphia: ASTM International.

Bachtiar, B., S. Sapuan, and M. Hamdan. 2008. "The Effect of Alkaline Treatment on Tensile Properties of Sugar Palm Fiber Reinforced Epoxy Composites." *Materials & Design* 29: 1285–90.

Baek, S., and Y. Kwon. 2007. "Optimization of the Pretreatment of Rice Straw Hemicellulosic Hydrolyzates for Microbial Production of Xylitol." *Biotechnology and Bioprocess Engineering* 12: 404–409.

Baker, J.M., P.J. Nixon, A.J. Majumdar, and H. Davies. 1991. *Durability of Building Materials and Composites*. U.K.: Longman Group Limited.

Bellis, M. 1997. "The History of Plastics." About.com Guide.

Brígida, A., A. Calado, and M. Coelho. 2010. "Effect of Chemical Treatments on Properties of Green Coconut Fiber." *Carbohydrate Polymer* 79: 832–38.

Castro, F.B. 1993. "The Potential of Dilute Acid Hydrolysis as a Treatment for Improving the Nutritional Quality of Industrial Lignocelluloses Byproducts." *Animal Feed Science and Technology* 42: 39–53.

El-Haggar, S.M. 2007. *Sustainable Industrial Design and Waste Management: Cradle to Cradle for Sustainable Development*. New York: Elsevier Academic Press.

Haque, M., and M. Hasan. 2009. "Physico-mechanical Properties of Chemically Treated Palm and Coir Fiber Reinforced Polypropylene." *Bioresource Technology* 20: 4903–4906.

Maya, J., and T. Sabu. 2004. "Mechanical Properties of Sisal/Oil Palm Hybrid Fiber Reinforced Natural Rubber Composites." *Composites Science and Technology* 64: 955–65.

Ngoc, H.P. 2006. "Production of Fibrous Activated Carbons from Natural Cellulose for Water Treatment Applications." *Carbon* 44: 2569–77.

Nkwachukwu, O., C. Chima, A. Ikenna, and L. Albert. 2013. "Focus on Potential Environmental Issues on Plastic World towards a Sustainable Plastic Recycling in Developing Countries." *International Journal of Industrial Chemistry* 4: 34.

Palmer, K., H. Sigman, and M. Walls. 2007. *The Cost of Reducing Municipal Solid Waste*. Washington, DC: Resources for the Future.

PEMRG (Plastics Europe Market Research Group). 2011. *An Analysis of European Plastics Production, Demand and Waste Data*. Belgium: Plastics Europe.

Pickering, K. 2008. *Properties and Performance of Natural-fiber Composites*. Cambridge, UK: Woodhead Publishing.

Seong, H., and S. Kim. 2008. "Application of Natural Fiber Reinforced Composites to Trenchless Rehabilitation of Underground Pipes." *Composite Structures* 86: 285–90.

USEPA (United States Environmental Protection Agency). 2006. *Municipal Solid Waste in the United States: 2005 Facts and Figures*. Official report, Municipal and Industrial Solid Waste Division. Washington, DC: US Environmental Protection Agency.

Van Den Einde, L., V.K. Karbhari, and F. Seible. 2007. "Seismic Performance of a FRP Bridge Pylon Connection." Invited paper for special issue of Composites: Part B on Structural Applications of Composites under Extreme Loadings. *Composites* 38: 685–702.

CHAPTER 6

Innovation in the Glass Industry: Upcycle of Glass Waste: Foam Glass

Dina Abdel Alim and Salah M. El-Haggar

Glass Industry

Glass is a non-crystalline, inorganic, nonmetallic ceramic material. It is made from inorganic materials that are fused at high temperature, then cooled to a rigid condition without crystallization, forming an amorphous structure. As a non-crystalline material, the glass molecules are not arranged in a repetitive long-range order. Its molecules change their orientation in a random manner (Smith, 1996). Glass, as a ceramic material, has a wide range of properties that make it indispensable for many engineering designs. It is hard and transparent at room temperature, along with excellent corrosion resistance to most substances in the normal working environment; it is also an electrical insulator. These properties make glass widely used in many fields, such as construction and the automotive industry, where it is used mainly as vehicle glazing. Tempered glass and laminated glass (safety glass) are used for automobile windows and windshields. The electronics and electrical industries use glass extensively because it is an insulating material that provides a vacuum-tight enclosure for electron tubes and lamps. High-tech glass is used for display screens and monitors in mobile phones, computers, and televisions. The chemical and pharmaceutical industries use glass because of its high chemical and corrosion resistance. Glass is used in laboratory apparatus and in liners for pipes and reaction vessels. In the household, glass is used for containers and tableware.

Table 6.1. Composition and properties of common commercial glasses (Smith, 1996)

Glass	SiO$_2$	Na$_2$O	K$_2$O	CaO
(Fused) silica	99.5+			
96% silica	96.3	< 0.2	< 0.2	
Soda-lime: plate glass	71–73	12–14		10–12
Lead silicate: electrical	63	7.6	6	0.3
High-lead	35		7.2	
Borosilicate: low expansion	80.5	3.8	0.4	
Low electrical loss	70		0.5	
Aluminoboro-silicate: standard apparatus	74.7	6.4	0.5	0.9
Low alkali (E-glass)	54.5	0.5		22
Aluminosilicate	57	1		5.5
Glass-ceramic	40–70			

B_2O_3	Al_2O_3	Other	Remarks
			Difficult to melt and fabricate but usable to 1000°C. Very low expansion and high thermal shock resistance.
2.9	0.4		Fabricate from relatively soft borosilicate glass; heat to separate SiO_2 and B_2O_3 phases; acid leach B_2O_3 phase; heat to consolidate pores.
	0.5–1.5	MgO, 1–4	Easily fabricated. Widely used in slightly varying grades, for windows, containers, and electric bulbs.
0.2	0.6	PbO, 21 MgO, 0.2	Readily melted and fabricated with good electrical properties. High lead absorbs x-rays; highly refractive, used in achromatic lenses and decorative crystal glass.
		PbO, 58	
12.9	2.2		Low expansion, good thermal shock resistance, and chemical stability. Widely used in chemical industry.
28	1.1	PbO, 1.2	Low dielectric loss.
9.6	5.6	B_2O, 2.2	Increased alumina, lower boric oxide improves chemical durability.
8.5	14.5		Widely used for fibers in glass resin composites.
4	20.5	MgO, 12	High-temperature strength, low expansion.
	10–35	MgO, 10–30 TiO$_2$, 7–15	Crystalline ceramic made by devitrifying glass. Easy fabrication (as glass), good properties. Various glasses and catalysts.

Most glass is constituted of a network of ionically covalently bonded silica (SiO_2) tetrahedra. Other oxides are added to glass to give it a range of properties that suit many applications. For example, the addition of Na_2O, K_2O, CaO, and MgO oxides to glass modifies the basic silica network and lowers the glass melt viscosity, which makes glass more workable and easier to form (Smith, 1996). Some of the common types of commercial glass are listed in table 6.1, along with their compositions and applications.

Soda-lime glass is the most widespread type of glass, representing approximately 90% of total production of glass. It is prepared by melting sodium carbonate (soda), limestone (lime), dolomite, silicon oxide (silica), aluminum oxide, and other fining agents at a temperature of around 1675°C. Soda-lime glass is used in applications such as containers, flat glass (used for windows), pressed and blown ware, and lighting products (where high heat resistance and chemical durability are not needed).

Additives such as Na_2O and CaO are added to soda lime to reduce the softening point of the glass from 1600 to 700°C to make it more workable. MgO is added to prevent devitrification (a condition in the firing process where glass develops wrinkles instead of a smooth glossy surface). Al_2O_3 is added in a small percentage to increase durability (Smith, 1996).

In lead glass, lead oxide is added so that the glass can be used as a shield against high-energy radiation. This property is important for applications such as fluorescent lamp envelopes, television bulbs, and radiation windows. High-lead glasses also have high refractive indexes, which make them good candidates for some optical glasses (Smith, 1996).

Permanent dyes are added to glass in order to give it different colors. Colors in glass are made by adding coloring ions and colloids. Metals and metal oxides are added during the manufacturing process to give it specific colors. For example, ferrous oxide is added to glass to give it a bluish-green color. Adding chromium with iron oxide gives glass a green color.

Viscous deformation of glass

The solidification behavior of glass differs from that of a crystalline solid. Figure 6.1 shows the solidification behavior of a crystalline vs. a non-crystalline material, where T_g is the glass transition temperature and T_m

is the crystalline material melting temperature. As the crystalline solid (for example, a pure metal) cools down, it crystallizes at its melting temperature with a significant decrease in its specific volume as shown by path ABC (Smith, 1996). When glass cools down it follows a path like AD. The liquid glass becomes more viscous as it cools down; it transforms from a soft rubbery state to a rigid brittle state in a narrow temperature range. Figure 6.1 shows that the slope of the curve decreases slightly as the temperature decreases; the point of intersection of these two slopes is the glass transition temperature (T$_g$). Glass transition temperature is a transformation point in the glass melting process, and it is structure sensitive. It can be defined as the center of a range of temperature in which glass changes from being brittle to being viscous and soft (Smith, 1996). Glass acts as a supercooled liquid above the transition temperature. As the temperature increases above the glass transition temperature, the glass melt viscosity decreases and the viscous flow takes place with more ease. Figure 6.2 shows the effect of temperature on the viscosity of the common commercial types of glass (Smith, 1996).

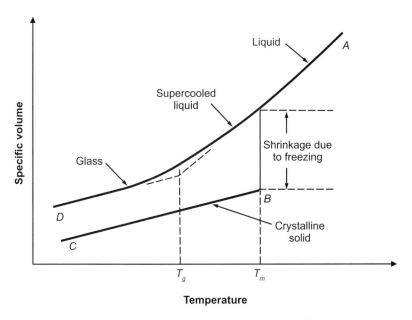

Figure 6.1. Solidification behavior of a crystalline vs. a noncrystalline glassy material (Smith, 1996)

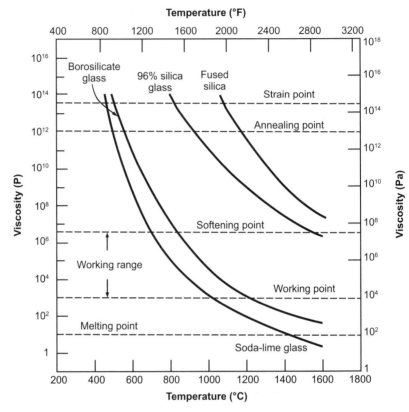

Figure 6.2. Effect of temperature on the viscosity of commercial types of glass
(Callister, 2000).

The horizontal lines denote several viscosity reference points. These points are (Smith, 1996):

- Working point: at this temperature the working of glass is easy.
- Softening point: at this temperature glass flows considerably under its own weight. This point cannot be defined precisely because it varies according to the glass surface tension and density.
- Annealing point: at this temperature the internal stresses of the glass are relieved.
- Strain point: below this temperature the glass is rigid and the stress relaxation process takes place at a very slow rate. The interval between the annealing and strain points is the annealing range of glass (Smith, 1996).

Glass manufacturing
Forming plate glass
Plate glass is formed by a process called the float process. The glass is melted in a furnace; then a ribbon of glass is moved out of the furnace and fed into a bath of molten tin, where it floats on its surface. Tin is suitable for the process because it is immiscible and has a high specific gravity. Glass cools down as it moves on the molten tin surface. When the surface is hard enough, the sheet of glass moves to an annealing furnace in order to remove the residual stresses, as shown in figure 6.3 (Smith, 1996).

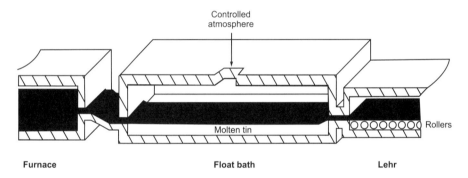

Figure 6.3. The float glass process (Smith, 1996)

Blowing, pressing, and casting of glass
In order to fabricate deep items like containers and light bulb envelopes, glass is shaped by blowing, in which a flow of air is let into the molten glass to force it into molds (Smith, 1996). The pressing method is used to fabricate flat items such as optical lenses. The molten glass is put into a mold; then a plunger presses the molten glass to take the shape of the mold (Smith, 1996) (fig. 6.4).

Casting is another forming technique for glass. In order to fabricate large simple-shaped objects like a large telescope mirror of 6m diameter, molten glass is cast into an open mold. Centrifugal casting is used for other types of objects that are hollow and have uniform wall thickness, like television tubes. The molten glass is poured into a spinning mold that makes glass flow to the wall of the mold, forming a glass layer of mostly uniform thickness (Smith, 1996).

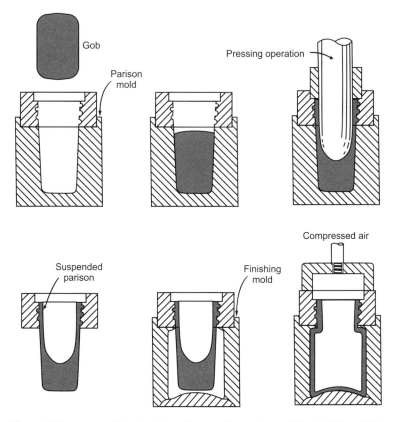

Figure 6.4. The press and blow technique for producing a glass container (Callister, 2000).

Glass production and recycling

Glass production is key in many other engineering sectors. Examples of different sectors that depend on glass are the construction, automotive, and electronic industries, and any other industry that uses container glass as a packaging material, such as the food industry in general. Due to rapid economic and technological development worldwide, there is a booming demand for glass products in various engineering fields.

According to Packaging News (2012), the production of container glass in the United Kingdom increased by 9.5%, from 2.12 million tons in 2009 to 2.32 million tons in 2010. Turkey showed the highest production rate increase of 27.2%, from 612.6 tons in 2009 to 779.5 tons in 2010. Figure 6.5 shows the market data of the European container glass production over a ten-year span from 2000 to 2010.

According to Glass on the Web (2012), the production of container glass increased in China by 18.55% from 2003 to 2004. The production in 2004 was 7.8522 million tons.

Flat glass is a key component in the construction and automotive industries. The rapid increase in these industrial sectors means increasing demand for flat glass. China is a fast-growing nation with robust economic growth. The construction industry is flourishing; China spends around 16% of its GDP on construction in order to accommodate its rapid urbanization (China Glass Industry Market Report, 2012). Approximately 2 billion square meters of construction floor space are completed every year (Glass on the Web, 2012). As for the automotive industry, China achieved, in 2010, 32% growth in light vehicles in comparison to other major countries (American Chamber of Commerce in Shanghai, 2012).

In order to meet the glass requirements of all the growing sectors, the production of flat glass is growing tremendously in China, increasing at an annual rate of over 20%, from 12.6 million tons in 2003 to 15.03 million tons in 2004.

The high speed of glass production means in turn a high speed of glass waste generation. Therefore a sustainable solution for glass waste is crucial. Recycling is the most sustainable solution for glass waste.

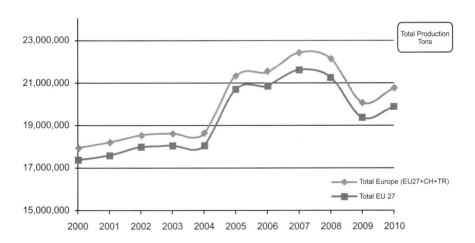

Figure 6.5. Total production of container glass in the European market in ten years (Packaging News, 2012)

Glass waste recycling

With a rising global population and increasing levels of consumption among developed and developing countries, pressure on earth's ecosystems will continue to increase in the foreseeable future. Societies are faced now with a real need to increase the efficiency of resources (material and energy), especially in developed countries. It is no longer accepted that raw materials are used for production and, after completing their life cycle, products are disposed of as waste to be landfilled or incinerated. The recycling of products at the end of their life cycle is becoming a crucial worldwide demand.

The amount of municipal solid wastes in Egypt is 28,000,000 tons per year. The percentage of glass waste is not unusually high with respect to other types of wastes; however, the weight percentage of glass waste is a significant figure. Table 6.2 shows the composition of the municipal solid waste in Egypt.

The US Environmental Protection Agency (EPA) reported that the amount of municipal solid waste (MSW) generated in the United States in 2010 was 250 million tons. Figure 6.6 shows the total amount of MSW classified by material. Glass waste represented 4.6% of the municipal solid waste. The weight of the municipal solid waste and the percentage of recovered material in the United States in 2010 are illustrated in table 6.3.

The United States generated 11.53 million tons of glass waste in the MSW stream in 2010, out of which only 27.1% was recovered.

Table 6.2. Composition of municipal solid waste in Egyptian cities (Solid Waste Management Strategy for Dakahleya Governorate, 1999)

Type of waste	Percentage
Organic waste	60
Paper	10
Plastic	12
Glass	3
Metals	2
Textiles	2
Other	11

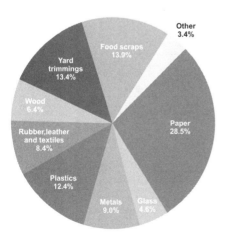

Figure 6.6. Total amount of MSW generated in the US in 2010 (Municipal Solid Waste Generation, 2011)

Table 6.3. Generation and recovery of MSW in the US in 2010, in millions of tons and percent of generation of each material (Municipal Solid Waste Generation, 2011)*

Material	Weight generated	Weight recovered	Recovery as percent of generation
Paper and paperboard	71.31	44.57	62.5
Glass	11.53	3.13	27.1
Metals			
Steel	16.9	5.71	33.8
Aluminum	3.41	0.68	19.9
Other nonferrous metals†	2.10	1.48	70.5
Total metals	**22.41**	**7.87**	**35.1**
Plastics	31.04	2.55	8.20
Rubber and leather	7.78	1.17	15.0
Textiles	13.12	1.97	15.0
Wood	15.88	2.30	14.5
Other materials	4.79	1.41	29.4
Total materials in products	**177.86**	**64.97**	**36.5**
Other wastes			
Food, other‡	34.76	0.97	2.8
Yard trimmings	33.40	19.20	57.5
Miscellaneous inorganic wastes	3.84	negligible	negligible
Total other wastes	**72.00**	**20.17**	**28.0**
Total municipal solid waste	249.86	85.14	34.1

* Includes waste from residential, commercial, and institutional sources
† Includes lead from lead-acid batteries
‡ Includes recovery of other MSW organics for composting
Details might not add to totals due to rounding
Negligible = Less than 5,000 tons or 0.05 percent

Table 6.4. Solid waste composition in the US and the UK as a percentage of weight (Solid Waste Management Strategy for Dakahleya Governorate, 1999)

Material	USA	UK
Food	7	20
Yard waste	18	4
Plastics	8	7
Glass	7	10
Metals	8	10
Miscellaneous	12	8
Other	—	8

The typical composition of solid wastes in the United States and the United Kingdom in weight percentage is shown in table 6.4 for comparison.

Glass containers represent about 6–8% of UK household waste; the recycling rate for the glass containers is approximately 30%, that is, almost 500,000 tons of cullet every year for industry and household wastes. In the United Kingdom, around 700,000 tons of flat glass for window and door glazing are produced yearly, 15% of which is recycled (Williams, 1998). Due to more organized and effective collection methods in most of the European countries, the glass recycling process is well established in Europe. Some European countries achieve recycling rates higher than 75% (Williams, 1998).

Classification of glass wastes
Fluorescent light bulbs

A fluorescent light bulb is defined as a "gas-discharge bulb that uses electricity to excite mercury vapor. The excited mercury atoms produce short-wave ultraviolet light that causes a phosphor to fluoresce, producing visible light. Mercury is an essential component of all fluorescent light bulbs, and allows these bulbs to be energy-efficient light sources" ("Fluorescent Light Bulbs," 2012). Because mercury is a hazardous material, fluorescent light bulbs are considered to be hazardous wastes. Manufacturers in the United States succeeded in reducing the amount of mercury used per fluorescent light bulb by 14% from 2001 to 2004. However, the total amount of mercury used in compact fluorescent light bulbs increased by 70% in the same period of time because of the increased amount of sales ("Fluorescent Light Bulbs," 2012). Each fluorescent light bulb contains from 10 to 100mg of mercury depending on its power capacity, while the compact fluorescent light bulb, which has less mercury per lamp, contains from 5 to 10mg of mercury. Since 2004, there has been a considerable increase in the production and use of electronic devices with fluorescent light bulbs for illumination in screen displays. Liquid-crystal display (LCD) screens have backlighting that contains fluorescent bulbs. These screens are lighted when the mercury is electrically energized. LCD screens are included in computers, flat-panel televisions, global positioning system (GPS) units, digital cameras, and handheld entertainment and communication systems.

Recycling of burned fluorescent light bulbs is the best way to prevent the mercury from being released into the environment through landfill and incinerators. This is why the US Environmental Protection Agency (EPA) is currently cooperating with the manufacturers and major retailers to develop and implement recycling opportunities ("Fluorescent Light Bulbs," 2012).

CRTs

A cathode ray tube (CRT) is the display component of the television and computer monitor. It contains between 15 and 90 pounds of glass. Some types of CRT contain 25% lead oxides, which are added to CRTs to protect the user from x-rays generated within the CRT during its operation ("CRT Glass to CRT Glass Recycling," 2012). Because lead is a hazardous material, CRTs should not be disposed of in trash or in municipal landfills, but in landfills for hazardous wastes. However, recycling CRTs is the best option for managing CRT waste.

Container glass wastes

Container glass is the glass used for making jars and bottles such as soft-drink bottles, beer bottles, mayonnaise and baby food jars, wine and liquor bottles, and many other food and beverage containers (Lund, 1993). Container glass is the only type of glass that is recycled in large quantities at the present time. Other types of glass, such as windowpanes, light bulbs, mirrors, glassware, crystal, ovenware, CRTs, and fiberglass, are not recyclable with container glass, as they are considered to be contaminants in container glass recycling (Lund, 1993). Container glass makes up over 90 percent of the amount of glass waste, while the remaining 10 percent comes from other types of glass (Lund, 1993). Permanent dyes are used to give different colors to container glass (figure 6.7). The most common colors are green, brown, and white (colorless). In industry, the green glass is called emerald, brown glass is called amber, and colorless glass is called flint (Lund, 1993). In order for recycled bottles and jars to meet certain manufacturing specifications, only amber or emerald cullet (crushed glass) can be used to make brown and green bot-

Figure 6.7. White, green, and brown glass containers (Viridor, 2012)

Figure 6.8. Public glass waste collection point in a neighborhood area for separating clear, green, and amber glass ("Glass Recycling," Wikipedia, 2012)

tles respectively (Lund, 1993). Thus a special consideration in collecting container glass is the need for color separation (figure 6.8).

A US company developed a method for overcoming the problem of permanent glass colors and the need for color separation. It covered the clear glass with a colored organic coating that melts away while the glass is recycled. This means that only one type of glass, clear glass, needs to be produced, since the color is no longer a constituent of the glass ingredients. This option extends the limit of the amount of cullet that might be recycled and simplifies the recycling process (Williams, 1998). Another solution is using container labels made of paper, foil, and plastic that must be removed during the recycling process. A photographic-quality ink was developed to label containers; it is applied directly to the glass (Williams, 1998).

Glass waste does not represent a threat to the environment: it is not hazardous because it is inert, and it is not biodegradable. When exposed to weathering forces, glass breaks down into small particles of silica (basic beach sand), which is one of the most common elements on earth (Lund, 1993). Glass recycling has grown considerably in recent years because the municipal solid waste collection is becoming more organized and glass manufacturers have increased their demands for recycled glass. Today, most glass manufacturers rely on a steady supply of cullet to supplement their raw material (Solid Waste Management Strategy for Dakahleya Governorate, 1999).

Container glass is made from common inert raw materials including silica sand, soda ash, and limestone. Slag, salt cake, feldspar, aragonite, and cullet are other ingredients that are used in container glass manufacturing. The mixture is heated to a temperature of 1675°C. The melted glass is shaped by a forming machine, where it is pressed or blown into the desired shape. The glass containers are left to cool slowly in an annealing furnace (Lund, 1993). Cullet from recycled container glass is always part of the recipe for new glass container production; sand is the only material used in greater volume than cullet in glass manufacturing (Solid Waste Management Strategy for Dakahleya Governorate, 1999). The amount of cullet used in glass containers differs from one manufacturer to another, but it is increasing. The average amount of cullet used is approximately 25 to 35% of the raw material needed for glass container manufacturing. Manufacturers expect to increase cullet use to 50%. However, bottle manufacturers have the capability to use up to 100% cullet. But this is virtually impossible in the production process, because it requires a consistent supply of cullet that must be completely clean, free of metals, and color sorted. It is the rigidity of the quality control process that reduces the percentage of cullet used by container manufacturers. The glass maker must be sure that the cullet used in the production process is of known type and color. In the production process, when a batch is started, chemicals must be added to compensate for impurities and color distortion of cullet (National Center for Resource Recovery, 1974). Glass manufacturers are conservative in specifying the percentage of cullet used because the supply of cullet is not steady (Lund, 1993).

The basic raw materials of glass (like silica sand, limestone, and sodium carbonate) are cheap; however, the energy needed to make glass is huge. Recycling glass will have a great advantage in reducing the energy needed to produce new glass products. Using cullet in new glass container manufacturing has several advantages.

- *Conserving energy and reducing manufacturing costs.* Energy is conserved because cullet melts down at a temperature lower than that required in melting the virgin raw material that is used in glass-making. This also increases the furnace life, which can be extended by as much as 15 to 20% depending on the amount of cullet being used. The conservation of energy will also conserve natural resources such as fossil fuel. It was found that for every 1% increase in the use of cullet, 0.25% of the needed energy is saved (Lund, 1993).

- *Reducing greenhouse-gas emissions.* Less energy needed by the furnace means reduced emissions of greenhouse gases, mainly CO_2.
  It was found that 315kg of CO_2 is saved for every ton of glass
  melted if recycled glass was used for new container manufacturing
  (after accounting for transportation and processing) (Waste
  Online, 2012). Moreover, the virgin material loses almost 15% of
  its input weight as waste gases, while no waste gases are released
  from cullet (Williams, 1998).
- *Reducing demand for raw material.* The raw materials for glass are
  abundant in nature but must be extracted from our landscape. It
  has been calculated that 1.2 tons of raw materials are preserved
  for every ton of recycled glass used (Waste Online, 2012).
- *Reducing the amount of waste glass that needs to be landfilled.*
  Although glass is inert and it does not represent a direct threat to
  the environment, losing it as a possible input for other industries
  and making it occupy large spaces in landfills is a waste of
  resources and is not a sustainable solution.

Container glass is unique in the recyclables manufacturing industry.
A 12-ounce glass bottle, melted down and reformed, yields a 12-ounce
bottle with the same quality and with no by-products. Furthermore, the
same glass can be recycled repeatedly while retaining its strength (Lund,
1993). This feature makes glass 100 percent recyclable; that means glass
should never reach a landfill.

Figure 6.9. Mix of green and white glass containers in a drop-off area in the
al-Gouna recycling center ("Industrial Ecology," 2007)

Recycling glass containers

Glass containers are usually recovered from drop-off centers that receive solid wastes. Figure 6.9 shows a drop-off center in the al-Gouna recycling center located near the al-Gouna tourist village in Egypt. Glass containers are separated either at the source or in the center upon receipt (Lund, 1993).

The processing of glass containers to cullet may or may not take place in the drop-off center. The glass container recycling process consists of:

1. Manual sorting: Glass containers are separated by color (green, brown, and mixed). Workers remove any large objects such as ceramic, stone, and plastic as the glass passes on a conveyor belt (fig. 6.10).
2. Crushing and screening: Glass containers are crushed by steel rollers to small pieces between 10 and 50mm. The crushed glass is screened by a vibrating bar screen that removes corks, papers, and any other larger items.
3. Ferrous metal removal: The screened cullet passes over a large magnet that rotates on a drum whose main function is to remove any ferrous metals such as bottle caps, wires, and steel.
4. Stone and ceramic removal: Cullet passes through another screen with a mesh base (about 19mm^2). Small stone and ceramic items pass through the mesh.
5. Vacuuming: Cullet passes through a strong vacuum duct that sucks out any paper, such as bottle labels.
6. Non-ferrous metal removal: Metals such as aluminum and lead that come from the bottle lids are removed. Cullet drops over a metal-detector head that directs a strong air jet to the non-ferrous metals, directing them away from the cullet stream side.
7. Small ceramic and stone removal: A laser beam is directed to the cullet. The beam is able to pass through the glass, but when it hits a piece of ceramic or stone it is reflected. When the laser beam is reflected, a computer fires an air jet that directs the stone and ceramic away from the cullet stream through a split chute that is divided into cullet stream and waste stream.

Figure 6.10. Glass bottles collected manually from the conveyor belt
("Industrial Ecology," 2007)

Some of the processing steps may take place at the manufacturer's site, or all the steps might take place at recycling centers that sell high-quality cullet.

Any foreign materials left with cullet, such as metals, ceramics, and stones, reduce the quality of the cullet. These foreign objects do not melt in the furnace with the glass, and they form bubbles and stones in the glass bottles that weaken their walls. It is impossible to remove every single bit of foreign material from cullet, but glass manufacturers can accept a minimum amount of foreign materials. One glass recycling company in the United Kingdom that sells cullet states that their quality control allows only 20g of foreign materials per ton of cullet (Glass Recycling UK, 2007).

Breaking glass before separation is an undesirable step because, as mentioned earlier, in order to meet certain manufacturing specifications, only amber or emerald cullet can be used to make brown and green bottles respectively. So the broken glass that cannot be color-sorted from the waste stream is of no value for glass container manufacturers. This type of mixed glass might be used in other applications, for example as a component of a composted waste product. Glass has the same physical properties as sand, which is why glass particles can be used in compost (Lund, 1993). Another use for recycled glass that does not meet the manufacturing specifications is to use it as aggregate for glassphalt. Glassphalt is the same as conventional asphalt except that some of the aggregates are replaced with crushed recycled glass ("Preparation and Placement of Glassphalt," 2012). Recent research studies have proved that mixed cullet can be used to produce foam glass, which is an insulating material with excellent properties (Eidukyavichus et al., 2004).

Other uses for glass cullet

Although glass container manufacturing is the primary market for glass cullet, it can also be used in other industries (Lund, 1993) such as:

- Soil conditioner, where glass is used to improve drainage and moisture distribution
- Glassphalt for road pavement, where crushed glass can be substituted for a percentage of the aggregates
- Glass wool as an insulating material
- Reflective paints used in road signs
- Filler for outdoor paint
- Artificial sand for beaches
- Fiberglass
- Abrasives. For example, in sandblasting operations, ground waste glass can safely replace silica sand.
- Lightweight aggregates in concrete
- Glass polymer composites. Finely ground glass is used as an additive in plastic. The glass filler increases the strength and ductility of plastics.
- Construction materials such as tiles, clay bricks, and glascrete
- Foam glass for insulation and construction products

Foam Glass

Foam glass is a lightweight glass material of cellular structure. It can be made entirely out of recycled glass. It is a valuable product that can be used in a range of engineering applications because of its structure. It is a heterophase system that consists of a solid phase and a gaseous phase. The solid phase is glass, which constitutes the thin walls of the cells; the wall thickness is several micrometers. Inside the cells is the gaseous phase. Foam glass has a very low density ($120kg/m^3$), yet a relatively high compressive strength and dimensional stability (Foamglas website). These characteristics make foam glass a good candidate for thermally and acoustically insulating construction materials.

The foam glass consists of millions of sealed glass cells, where every cell represents an insulating space. This makes it a highly effective insulating material. The closed-cell structure makes the material watertight since no diffusion can take place; this makes foam glass an efficient barrier against soil humidity (Foamglas website). Most importantly, foam glass can be manufactured fully out of waste glass with minimum addition of virgin materials.

Foam glass has many applications. The most prevalent is as a heat-insulating material. Its service parameters as a heat insulator surpass many other organic types of foam. It has a relatively high mechanical strength that facilitates its installation. It is resistant to water, moisture impermeable, incombustible, and biologically resistant because it does not putrefy. In addition, it is chemically inert. All these properties ensure the constancy of the thermal conductivity value of foam glass over time (Foamglas website). Foam glass insulation is an environmentally sustainable solution for recycling glass wastes. Figure 6.11 explains the foam glass processing technique adopted by Pittsburgh Corning Company for foam glass insulation, as explained below (Foamglas website).

1. Mixing of the raw materials: silica sand and other additives are melted to produce glass. Glass waste is added, up to 60%.
2. Raw materials are fed into a furnace with a constant temperature of 1250°C.
3. Molten glass is drawn out of the furnace onto a conveyor belt.
4. Control room.
5. The molten glass cools down as it moves onto the conveyor belt; then it is crushed before being fed to the ball mill.
6. Carbon black is added.
7. Ball mill grinds and mixes crushed glass with the carbon black.
8. The fine powder mix is put into stainless steel molds that pass through a cellulating oven (sintering furnace) where the glass mix is foamed.
9. Heat energy recovery system.
10. The foam glass blocks are annealed in an annealing furnace to remove any residual stresses.

11. The foam glass blocks are cut and the production wastes return back to the process.
12. The foam glass products are packaged and labeled.
13. The foam glass products are stored and ready for transportation.

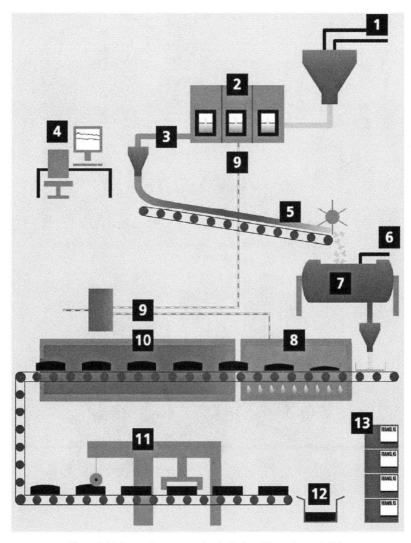

Figure 6.11. Foam glass processing technique (Foamglas website)

Figure 6.12 shows a block of foam glass when it comes out of the sintering furnace. Figure 6.13 shows the Pittsburgh Corning foam glass products; they include slabs, boards, elbows, and piping insulation segments.

Figure 6.12. Block of foam glass coming out of the sintering furnace
(Foamglas website)

Figure 6.13. Foam glass products (Foamglas website)

Foam glass has been known as a thermal insulation material since the middle of the twentieth century (Manevich and Subbotin, 2008). Interest in foam glass as insulation has been growing as fuel costs have increased. Heating residences is crucial for cold countries like those of North America and Western Europe. In Russia, 500–600 kWh/m^2 of living space is consumed for heating residences. Heat loss means energy loss. Effective thermal insulation will reduce fuel consumption for heating by a minimum of 25% (Manevich and Subbotin, 2008). It will also decrease the harmful emissions generated from burning fuel. In 2007, the volume of heat-insulating materials that were used in construction was 23–25 million m^3; at that time, it was estimated that this amount would increase to 45–50 million m^3 by 2010 (this amount includes industry and engineering systems) (Manevich and Subbotin, 2008). Table 6.5 summarizes the technical characteristics of different types of insulating materials, including foam glass. It shows that the commercial value of foam glass thermal conductivity is around 0.05–0.08 W/m.K. Foam glass has high compressive strength compared to other insulating material (> 0.7 MPa), higher water vapor impermeability, high maximum temperature of use (up to 450°C), and, most importantly, it is incombustible. Mineral wool and plastic foam, widely used in construction, are combustible; they contain more than 5% polymeric binders that release toxic gases when burned (Manevich and Subbotin, 2008). On the other hand, foam glass is a fire-resistant insulating material that is totally environmentally friendly.

The thermal resistance of plastic foams drops by 25% after six months. They can absorb water vapor that results in corrosion, degradation, and further decrease of thermal resistance (Manevich and Subbotin, 2008).

The service life of mineral wool and plastic foam is less than 12 years, which is less than the lifetime of most buildings and structures. However, foam glass has unlimited service life and it keeps its characteristics throughout its lifetime. The excellent service parameters make foam glass an effective insulating material in many applications.

Applications of foam glass
Construction applications for foam glass thermal insulation
Foam glass is used for insulating walls, roofs, and floors of buildings in both cold and hot regions (foam glass is also a sound-insulating material) (Manevich and Subbotin, 2008). Figure 6.14 shows a layer of foam glass in a concrete roof deck. The foam glass slab is fixed on the primed concrete by hot bitumen. It forms a secondary waterproof layer below the asphalt layer.

Table 6.5. Technical characteristics of different types of heat-insulating materials
(Manevich and Subbotin, 2008)

Materials	Density, kg/m³	Compressive strength, KPa	Thermal conductivity, W/(m·K)
Mineral wool products			
Heat-insulating tiles made of mineral wool on synthetic binders	200	40–120	0.060–0.066
High-rigidity mineral-wool tiles	200	100	0.045
Glass staple fiber tiles	175–200	—	0.049
Foam plastics			
Polystyrene foam	20–40	50–150	≈ 0.038
Polyurethane foam	40–60	120–200	0.035
Heat-insulating tiles made of reol resins	50–100	50–200	0.040–0.046
Cellular inorganic materials (foam concrete)	350	700–800	0.090–0.100
Foam glass			
Blocks	120–200	≥ 700	0.050–0.080
Gravel	80–100*	500–1000	0.030–0.050
Keramzit (gravel)	250–350	500–1500	0.210–0.230

* Specific bulk mass

Maximum temperature of use, °C	Water vapor permeability, mg/(m·h·Pa)	Combustibility	Environmental evaluation	Lifetime, years
100	0.380–0.600	Difficult	Toxic when burned (7–8% binder)	7–10
—	0.58	Difficult	Toxic when burned (10% binder) and inhaled	10–12
180	—	Difficult	Toxic when burned (6% binder) and inhaled	7–10
70	0.05	Combustible	Toxic when burned	10–12
200	—	Combustible	Toxic when burned	10–12
130	—	Combustible	Harmful emissions	10–12
350	0.23	Incombustible	Environmentally clean	~10 (when protected from moisture)
450	0.001–0.004	Incombustible	Environmentally clean	Unlimited
450	—	Incombustible	Environmentally clean	Unlimited
450	0.21	Incombustible	Environmentally clean	Long-lived

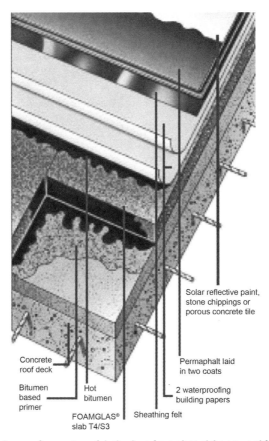

Solar reflective paint,
stone chippings or
porous concrete tile

Concrete
roof deck

Permaphalt laid
in two coats

Bitumen
based
primer

Hot
bitumen

2 waterproofing
building papers

FOAMGLAS®
slab T4/S3

Sheathing felt

**Figure 6.14. Layers of concrete roof deck where foam glass slabs are used for insulation
(Foamglas website)**

Figure 6.15 shows an internal wall insulated with foam glass. Foam glass reduces the effect of thermal bridges and prevents condensation damage that affects the interior finish. Foam glass boards are fixed to the wall by adhesives and mechanical fixings.

Industrial applications for foam glass thermal insulation

Foam glass is used for insulating systems where the operating temperature is below ambient, as well as in cryogenic process systems. It is also used for insulating hot systems. The operating temperature ranges from -268°C to 482°C. Examples are industrial refrigerators and ships, low-temperature pipelines, chilled water lines, low-temperature storage

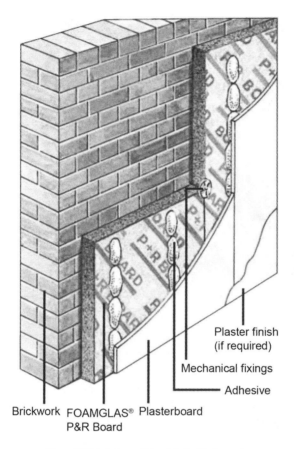

Plaster finish
(if required)

Mechanical fixings

Adhesive

Brickwork   FOAMGLAS®   Plasterboard
                P&R Board

**Figure 6.15. Interior wall insulated with foam glass
(Foamglas website)**

spheres, high-temperature pipelines and equipment, underground piping, and vertical storage tank walls and roofs (Foamglas website).

Foam glass is suitable for underground applications because it is strong, and resistant to moisture and corrosion. Being inorganic, it resists vermin, insects, and microorganisms, and it does not putrefy. Figure 6.16 shows the installation of two layers of foam glass insulation around a liquefied natural gas (LNG) pipe in the Isle of Grain LNG phase 2 project in the United Kingdom. The project had a strategic national importance; therefore a reliable insulating material was required. Foam glass was chosen because it can be installed rapidly with more flexibility (Foamglas website).

Figure 6.16. LNG pipe insulated with two layers of foam glass
(Foamglas website)

Another possible application of foam glass insulation is in atomic energy applications where fire safety is imperative (Manevich and Subbotin, 2008). Table 6.6 summarizes the production volumes and prices for heat-insulating materials in different countries. The price of 1m³ of foam glass ranges between $150 and $500 depending on the quality of the foam glass and the supply and demand. PittsburghCorning Europe (Belgium) is the largest foam glass company in Europe, producing 860,000m³/year (Manevich and Subbotin, 2008).

Other applications for foam glass
Other than thermal insulation, foam glass can be used for many applications, such as lightweight gravel and filling material. The foam glass gravel might be the scrap generated during the foam glass block cutting process, or it might be produced through a separate process. In the latter case the annealing step is not required (Manevich and Subbotin, 2008). Foam glass can be used as low-weight filling material in construction projects (fig. 6.17). It was used in road fill over a tunnel along one of the major roadways in Norway (Reindl, 2003). It can also be used in filling trenches around embedded pipes ("Foamed Glass Technology," n.d.).

Table 6.6. Production volumes and prices for heat-insulating materials in different countries (Manevich and Subbotin, 2008)

Country	Type of heat insulation	Production volume per 1000 inhabitants, m³	Price of 1m³, USD
Russia	All types	53	47–54* 78–120**
	Foam glass only	0.05	120–140 300–400† 500‡
Belarus	Foam glass	48	100
Japan	All types	350	—
	Foam glass only	50	300
Finland	All types	400	—
USA	All types	500	—
	Foam glass only	120	300–400
Germany	All types	70	—
Sweden	All types	600	—
Belgium	All types	350	—
	Foam glass only	85	150–200§

* Polystyrene foam
** Mineral wool tiles
† In Moscow
‡ In Siberia
§ In Europe

Foam glass blocks can be used as an abrasive material, such as in stone-washing jeans. They can also be used for coating removal and surface preparation (Reindl, 2003). Other functions include acting as a biological filter. A biological filter consists of a bed of rock, gravel, or foam media over which wastewater flows downward, causing a layer of microbial slime. The organic matter in the wastewater is adsorbed and absorbed by the layer of microbial slime that covers the media bed. The

Figure 6.17. Spreading lightweight foam glass as a filling material
(Miracle Sol Association website)

diffusion of the wastewater over the media is important in order to provide the oxygen that the microbial slime layer needs for the biochemical
oxidation of the organic matter. Foam glass, as a highly porous material,
has a large surface area that is very efficient when used as a bed medium
for water treatment (Miracle Sol Association, 2012).

Foam glass production is affected by many processing parameters,
among them sintering temperature, or the temperature at which the glass
powder/foaming agent mix is transformed into foam glass; the length of
the foaming agent soaking time, which is the time that the mix is left in
the furnace, at the sintering temperature, in order to be transformed into
foam; and glass powder particle size. The color of the glass certainly has
an effect on the properties of the foam glass: different glass color means
different starting materials, since they have different compositions.

Techniques of producing foam glass

Foam glass can be produced through two different processes. The first
consists of blowing gases into the molten glass. The second consists of
foaming sintered fine glass powder by adding a foaming agent. The
glass powder:foaming agent ratio percentage (by weight) is usually

optimum in the range of 97–99.5:1.5–3 (Brusatin, Bernardo, and Scarinci, 2004; Fernandes, Tulyaganov, and Ferreira, 2009; Liaudis et al., 2009; Wu et al., 2006). Foaming agents are additives that combine to chemically react and release gases, or that decompose when heated. The main role of the additive is to create the gaseous phase during heating; it is added to the glass powder in small quantities. The additive is called pore-forming, gas-forming, blowing agent, or foaming agent. Currently, most foam glass production is done using the second method, the foaming agent technique (Brusatin, Bernardo, and Scarinci, 2004; Fernandes, Tulyaganov, and Ferreira, 2009; Liaudis et al., 2009; Wu et al., 2006; Mear et al., 2006; Bernardo et al., 2007; Mear et al., 2007). The foaming agent technique is much less expensive and requires less sophisticated technology.

To produce foam glass through the foaming agent technique, the glass powder and foaming agent must be heated to a temperature above the glass softening point. The sintering temperature is usually in the range of 800–1000°C. During the thermal treatment of the glass powder and additives mixture, several processes take place. When the temperature reaches the softening temperature of glass, the glass powder starts to sinter and form a continuous sintered body. At this stage, the particles of the gas-forming agents are insulated by the softened glass. They start to emit gases after a certain temperature, frothing the melted glass (Spiridonov and Orlova, 2003). The emission of gases creates the insulated pores throughout the glass melt; this occurs where the particles of the gas-forming agents are blocked inside the softened glass. The foaming reaction must start only after the glass particles are sintered, otherwise the gas will escape from the glass powder. The gas evolution due to the chemical reaction between the glass powder and the foaming agent creates bubbles inside the glass melt. The bubbles grow due to the increase of the gas pressure inside the bubbles. This is caused by the chemical reaction that generates gases and by the increase of the sintering temperature inside the furnace, as illustrated by figure 6.18.

As the temperature increases, the surface tension of the glass melt decreases. Consequently, the external pressure exerted on the bubbles (by surface tension) decreases, enhancing the bubble growth. The growth rate of the bubbles depends mainly on the glass melt viscosity and surface tension. The viscosity acts as a resistance against bubble growth (Seiner, 2006).

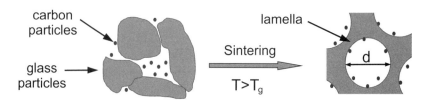

Figure 6.18. Schematic representation of the bubble formation and growth in a glass/carbon powder mixture upon sintering to produce foam glass (Seiner, 2006).

Types of foaming agents

There are two types of foaming agent: neutralization and redox. Neutralization agents include salts (carbonates or sulfates) that decompose when heated, emitting CO_2. The gas release is so intense it breaks the walls of the isolated pores during its decomposition, making the internal structure look like a maze system of merged cavities. This type of foam glass has high water absorption and soundproof parameters. The neutralization foaming agent ensures high water absorption of approximately 50–70% (Seiner, 2006). On the other hand, the redox agents—such as carbon-containing materials like coke, soot, anthracite, graphite, carbon black, and, less frequently, silicon carbide—go through an oxidation reaction with the gases dissolved in the glass melt, such as oxygen and sulfur anhydride. The oxidation process releases gases that produce foam glass, inside which sealed pores prevail. This type of foam glass is mainly used as a heat-insulating material. Redox foaming agent has a lower water absorption of approximately 10–15% (Seiner, 2006). Foam glass with sealed pores has lower thermal conductivity than foam glass with communicating pores. Thermal conductivity is affected by the type of porosity: its shape, its size, and its distribution inside the foam glass. It depends mainly on the ratio between the solid and the gaseous phases. The pores inside foam glass, which represent the gaseous phase, exceed 90% and gases in general have low thermal conductivity. This is why foam glass has low thermal conductivity in general and lower thermal conductivity values when its pores are sealed (Seiner, 2006).

Carbonaceous materials have been used as foaming agents in various studies (Manevich and Subbotin, 2008; Wu et al., 2006; Mear et al., 2006; Bernardo et al., 2007; Mear et al., 2007). Bernardo et al. (2007)

used commercial carbon black as a foaming agent with the amount of 0.5 wt % at a sintering temperature of 850°C for 30 minutes. The foam glass produced had closed porosity (around 95%) and a density lower than 0.3 g/cm^3. Silicon carbide (SiC) is an effective foaming agent that has been used by many researchers (Wu et al., 2006; Mear et al., 2006; Mear et al., 2007). Wu et al. (2006) produced foam glass with uniform pore morphology by adding 2 wt % SiC and heat treatment at 1000°C for 5 minutes. The compressive strength of the foam glass thus produced was in the range of 1.2–1.7 MPa. The only drawback of SiC is that it decomposes at high temperature (950–1150°C), and thus increases the production cost despite the short time (5 minutes).

Carbonates have been used as foaming agents in various studies (Fernandes, Tulyaganov, and Ferreira, 2009; Bernardo et al., 2007). For example, Bernardo et al. (2007) used calcium carbonate (CaCO$_3$) as a foaming agent. The decomposition of CaCO$_3$ leads to extensive foaming, resulting in open cell morphology. CaCO$_3$ usually agglomerates when mixed with glass powder, so that the mixture needs long mixing (30 minutes) to achieve homogeneous distribution of pores. Fernandes et al. (2009) used calcium magnesium carbonate (also called dolomite, CaMg(CO$_3$)$_2$) and calcite-based sludge derived from the marble cutting process (consists of about 99% CaCO$_3$) as foaming agents. The foams that comprised 2 wt % CaMg(CO$_3$)$_2$ and 2 wt % marble waste (separate samples) had low densities (around 0.36 g/cm^3) and compressive strength (around 2.4 MPa) at sintering temperature = 850°C. Fernandes et al. (2009) achieved better mechanical properties (compressive strength around 2.4 MPa) at low sintering temperature (850°C) than Wu et al. (2006), who used SiC as a foaming agent, with lower mechanical properties (maximum compressive strength 1.7 MPa) at much higher temperature (1000°C).

Producing foam glass from recycled glass

Foam glass can be produced out of recycled glass, either from a mixture of waste glass with different chemical composition or from one type of waste glass. It can also be produced fully out of recycled glass or part recycled glass and part virgin material. The utilization of glass waste in most industrial applications requires a strict level of purity. To produce conventional (container) glass, only cullet of well-controlled composition is acceptable. Thus, use of glass waste in engineering applications where no strict requirement of purity is needed allows 100% recycling of glass wastes. Foam glass production is an exemplary process for fully recycling large amounts of waste glass to produce high-quality foam. Eidukyavichus et al. (2004) used cullet of different chemical composition (mixture of window and container glass) to produce foam glass; Wu et al. (2006) used a mixture of 80 wt. % bottle glass cullet and 20 wt. % coal pond ash; Liaudis et al. (2009) used recycled flat glass; and Mear et al. (2006) used CRT wastes. Manevich and Subbotin (2008) state that the initial glass for producing foam glass can be in-house or purchased cullet or a mixture of both. The initial raw material can be vertically drawn sheet glass, float glass, rolled glass, colorless bottle glass, or green glass. While the use of brown containers, electric lamps, and lead crystal requires additional processing, sheet and container glass (colorless and green) can be added to the initial material to achieve high-quality, low-cost foam glass (Manevich and Subbotin, 2008).

Properties of recently developed foam glass

The Pittsburgh Corning foam glass company (Foamglas website) determines the properties of foam glass for high-load-bearing applications such that the average density, for all grades, is 0.12–$0.16 g/cm^3$, the compressive strength is 0.8–1.6 MPa, and thermal conductivity at $10°C$ is 0.043–0.048 W/(m.K).

The foam glass that was recently developed in laboratories on a research scale has a density in the range of 0.2–0.4 g/cm^3 and compressive strength of 0.5–3 MPa (Fernandes, Tulyaganov, and Ferreira, 2009; Liaudis et al., 2009; Wu et al., 2006). The typical range of thermal conductivity of recently produced foam glass is 0.05–0.08W/(m.K) (Manevich and Subbotin, 2008). Table 6.7 shows the properties of some recently produced foam glass samples.

Case study: Producing high-quality foam glass from container glass waste at the American University in Cairo labs

Foam glass with excellent properties has been produced at the American University in Cairo's mechanical engineering labs. Abdel Alim (2009) used recycled soda-lime glass to produce foam glass with high quality (fig. 6.19). It had low thermal conductivity (0.078 W/m.°C) with low bulk density ($0.25 g/cm^3$) and high porosity (90%). These properties are excellent for thermal insulation applications because the thermal conductivity of the foam is low; it also has good compressive strength (1.62 MPa) compared to other insulating material.

Table 6.7. Properties of foam glass samples (Shutov et al., 2007)

Sample no.*	Foaming temperature, °C	Foaming agent	Mass content, %	Density, kg/m³	Compressive strength, MPa	Porosity, %	Average pore size, mm
1	910	chalk	2.5	373	2.7	85.0	1.05
2	910	chalk	2.5	340	2.9	86.4	1.15
3	910	chalk	2.5	152	1.9	94.0	3.10
4**	No data			100	1.1	96.0	0.58
5	910	chalk	2.5	241	2.7	90.0	1.36
6	790	carbon, carbon black	0.9	210	2.0	92.0	0.82
7	910	chalk	2.5	395	2.1	84.0	No data
8	900	carbon black	0.2	329	3.1	86.7	0.19
9	910	chalk	2.5	238	2.2	90.5	0.30

* In all cases, the specific surface area of the batch was 400 m²/kg
** Sample from Foamglas Co.

**Figure 6.19. Cubic sample of foam glass produced
from recycled container glass**

Material preparation and experimental procedure

The starting material was waste container glass from MSW. Sodium silicate solution ($Na_2SiO_3 \cdot 9H_2O$) in the form of water glass was used as a foaming agent. The waste container glass from MSW is always contaminated and needs some preparation steps before it can be used in experimentation. The container glass wastes with different colors were cleaned from contamination by a washing machine, dried in air, crushed, and ground in a ball mill. Then the glass powder was sieved into 3 particle sizes: size 1 = 75μm, size 2 = 150μm, size 3 = 250μm. Figure 6.20 shows the process flow diagram of the container glass waste preparation.

The glass powder was mixed with the foaming agent at different weight percentages. The mixture of glass powder and sodium silicate solution was then mixed for 5 minutes in a rotary mixer. Figure 6.21a and b shows the glass powder and the wet glass powder/sodium silicate solution mixture. The mixture was then cold-compacted at 10 tons using a uniaxial laboratory hydraulic press for 20 minutes (fig. 6.21c). The slabs were dried in an oven for 1 hour at 200°C. The dried slabs were sintered in an electric laboratory furnace; they were put in directly at temperature in the range of 750–920°C. After completing the required soaking time, samples were severely cooled by natural convection at a rate of 40°C/min until the temperature reached 600°C in order to freeze the evolution of

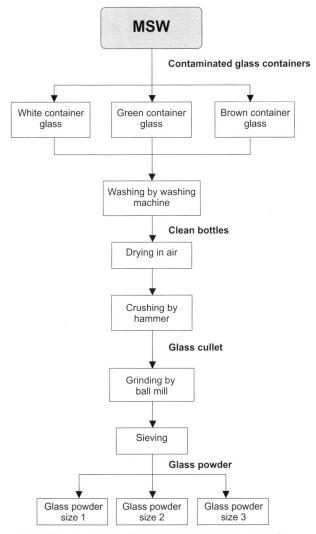

Figure 6.20. Process flow diagram for container glass preparation

the microstructure. Then the foam glass slabs were cooled slowly from 600°C to 500°C at a rate of 1°C/min to be annealed at a temperature close to, but slightly lower than, the softening temperature of soda lime glass (≈ 700°C). They were then cooled to room temperature at a rate of 25°C/min (fig. 6.21d shows a foam glass slab after annealing). Figure 6.23 illustrates the processing steps for foam glass production.

Figure 6.21. (a) Glass powder; (b) glass powder/sodium silicate solution mix; (c) pressed slab prior to drying; (d) foam glass slab

The foam glass samples were machined using a bench-type circular saw. Figure 6.22 shows foam glass pieces of different sizes cut with high precision.

Figure 6.22. Foam glass samples cut with high precision

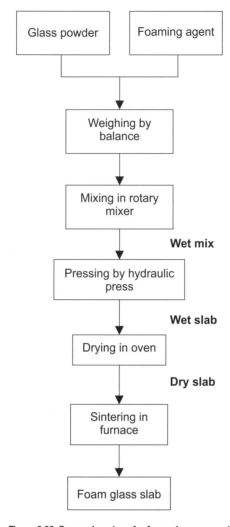

Figure 6.23. Processing steps for foam glass preparation

Results

The physical and mechanical properties of the produced foam glass were studied. Density, compressive strength, thermal conductivity, and pore morphology were analyzed for different experimental variables such as sintering temperature, amount of foaming agent, soaking time, glass powder particle size, and glass powder color.

Influence of sintering temperature

The influence of sintering temperature on foam glass was studied by subjecting compositions that contained 12 wt. % sodium silicate solution to sintering temperatures that ranged from 750 to 900°C in increments of 50°C (using white glass with particle size 75μm, soaking time 30 minutes).

The results showed that the sintering temperature has a considerable effect on the bulk density and the degree of foaming achieved in the produced foam glass. Generally, as the sintering temperature increased from 750°C to 900°C, the bulk density decreased while the percentage of porosity increased, as shown in figures 6.24 and 6.25.

The bulk density of the foam decreased continuously as the sintering temperature increased. The foam glass has a relatively high bulk density (ϱ_b = 0.61 g/cm^3) at 750°C; its relative density (ϱ_r = 0.24) is somewhat high for a foam. Gibson and Ashby (1999) state that foams must have relative densities less than 0.3. The bulk density dropped by approximately 31% from 750°C to 800°C (0.61 to 0.42 g/cm^3). A considerable further drop of bulk density (approximately 40% from 0.42 to 0.25 g/cm^3) was achieved as the temperature increased from 800°C to 850°C. As the temperature increased from 850°C to 900°C, the bulk density dropped by only around 12% (from 0.25 to 0.22 g/cm^3).

Figure 6.24 shows that the compressive strength decreased as the sintering temperature increased. The composition that was sintered at 750°C had the highest value of compressive strength (18.68 MPa). This was expected because it has a lower amount of foaming agent and low sintering temperature; consequently, the foaming of the sample was relatively low (it has a porosity of 76%). The compressive strength dropped tremendously, by around 76% (from 18.68 to 4.43 MPa), at a sintering temperature of 750 to 800°C (as the percentage of porosity increased from 76 to 83%). A further drop occurred from 800 to 850°C by around 63% (from 4.43 to 1.62 MPa). The minimum compressive strength achieved equaled 0.81 MPa; it corresponds to 900°C and ϱ_b = 0.22 g/cm^3.

The morphological evolution of foam glass (fig. 6.25) shows that as the sintering temperature increases, the foaming process inside the material increases. As the temperature increases, the viscosity of glass melt decreases and the pressure of the gas entrapped inside the pores increases as well. Consequently, the pore size increases as the sintering temperature increases (Fernandes, Tulyaganov, and Ferreira, 2009).

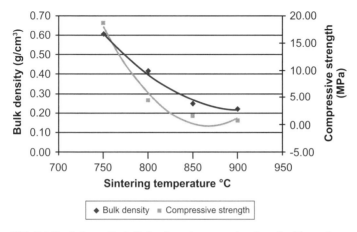

Figure 6.24. Relation between the bulk density and compressive strength of foam glass with 12 wt. % sodium silicate solution as a function of the sintering temperature

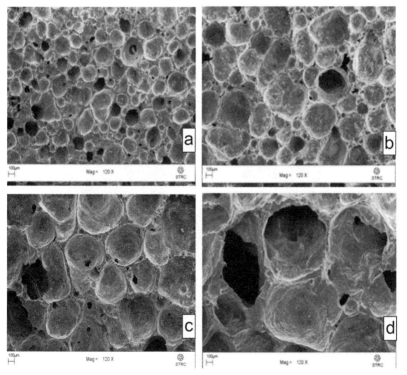

Figure 6.25. High magnification of the morphological evolution of foam glass with 12 wt. % sodium silicate solution at sintering temperature equal to (a) 750°C, (b) 800°C, (c) 850°C, (d) 900°C.

The most homogeneous structure is found in the foam glass sintered at 850°C; below 850°C and beyond 850°C, the pore size distribution is comparatively inhomogeneous. As the temperature increases, pores tend to combine, forming larger pores, as previously explained. This results in a coarser structure with larger pore size and inhomogeneous distribution of pores. The morphology of the foam glass sintered at 900°C shows that the coalescence phenomenon took place extensively, forming larger interconnected pores (see white circles in figure 6.26 showing coalescence of pores).

The thermal conductivity of the foam glass samples was measured. Figure 6.27 plots the variation of thermal conductivity and percentage of porosity as a function of the sintering temperature.

Materials that have thermal conductivity less than 0.25 W/m.°C are classified as insulating materials (Mear et al., 2007). All the measured samples had thermal conductivity lower than 0.25 W/m.°C. The thermal conductivity was found to vary from 0.092 W/m.°C (foam glass sintered at 750°C) to 0.053 W/m.°C (foam glass sintered at 900°C), which is close to the value of the thermal conductivity of the newly developed foam glass (0.05–0.08W/m.K) (Manevich and Subbotin, 2008). It is slightly higher than the thermal conductivity of the commercial foam glass produced by Pittsburgh Corning, which equals 0.043–0.048 W/m.K (Foamglas website). Thermal conductivity was

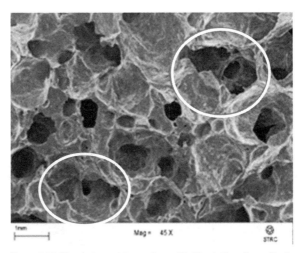

Figure 6.26. Morphology of foam glass with 12 wt. % sodium silicate solution at sintering temperature 900°C

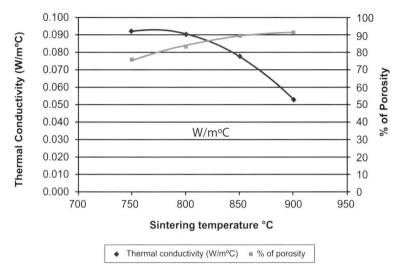

W/m°C

Figure 6.27. Variation of thermal conductivity and percentage of porosity as functions of the sintering temperature for foam glass with 12 wt. % sodium silicate solution

found to decrease with increasing porosity percentage. It reached the minimum (0.053 W/m.°C) at a 91% percentage of porosity. This relation between thermal conductivity and the percentage of porosity is expected because as the porosity increases inside the foam, its thermal conductivity tends to the thermal conductivity of air (thermal conductivity of air at room temperature is around 0.023 W/m.°C) (Mear et al., 2007). The values of the thermal conductivity of the first two compositions (at 750°C and 800°C) are approximately the same (0.092–0.090 W/m.°C). However, thermal conductivity dropped significantly from 800°C to 850°C (0.090–0.078 W/m.°C). A further drop took place at 900°C, to 0.053 W/m.°C.

The previous analysis of the bulk density, compressive strength, thermal conductivity, and morphology of the foam glass produced in this procedure shows that the sintering temperature has a great effect on the physical and mechanical characteristics of foam glass. By controlling the sintering temperature, the properties of the foam glass can be tailored to suit different applications. Foam glass with different combinations of service parameters can be produced at different sintering temperatures depending on the required characteristics. For applications that require foam with relatively high structural characteristics along with good thermal conductivity, the foam sintered at lower temperatures (750–800°C) at 12 wt. % sodium silicate solution will be more suitable, because it has high compressive strength (4.43–18.68 MPa) along with low thermal conductivity (around 0.09 W/m.°C). For thermal insulation applications, an excellent combination of properties was achieved at sintering temperature 850°C and sodium silicate solution amount 12 wt. % (ϱ_b = 0.25 g/cm^3, compressive strength = 1.62 MPa, thermal conductivity = 0.078 W/m.°C, percentage of porosity = 90%), along with a homogeneous morphology with a pore size around 0.5mm. The typical technical characteristics of foam glass for insulation applications range around 0.12–0.20g/cm^3, compressive strength > 0.7 MPa, and thermal conductivity 0.05–0.08 W/m.K (Manevich and Subbotin, 2008).

The highest sintering temperature (900°C) did not produce foam glass with satisfactory characteristics because, although the thermal conductivity of the foam was low (0.053 W/m.°C), the mechanical resistance of the foam was very low (0.81 MPa) and the morphology was inhomogeneous, full of coarse interconnected pores. Moreover, the density and percentage of porosity did not increase significantly from 850°C to 900°C (0.25–0.22 g/cm^3 and 90–91% respectively). Therefore, it is not recommended to use sintering temperature above 850°C because beyond this temperature, foam glass properties are not satisfactory. Most importantly, increasing the sintering temperature beyond 850°C will consume more energy and increase production cost.

The foam glass produced in this experiment had properties superior to the foam glass recently produced in laboratory research and at much lower sintering temperature. Brusatin, Bernardo, and Scarinci (2004) recycled foam glass from soda lime glass; they report that the foam glass produced had ϱ_b = 0.55–0.65 g/cm^3 and thermal conductivity = 0.124–0.136 W/m.K and that the optimum sintering temperature was 950°C. Wu et al. (2006) also recycled foam glass from soda lime glass, and found that the foam glass produced had ϱ_b = 0.2–0.4 g/cm^3 and compressive strength = 1.5 MPa, and that the optimum sintering temperature was in the range of 1000–1050°C.

It is not recommended to increase the sintering temperature to 900°C because at that temperature, the structure of the foam was coarse with large interconnected pores and was mechanically weak. The optimum sintering temperature is 850°C for preparing foam glass for thermal insulation.

Effect of the amount of foaming agent

The effect of the amount of foaming agent on the foam glass produced was studied by preparing compositions with 6, 12, 19, and 32 wt. % sodium silicate solution (using white glass with particle size 75µm, sintering temperature 850°C, and soaking time of 30 minutes). Figure 6.28 compares the foam prepared with different amounts of foaming agent in terms of bulk density, compressive strength, and morphology.

The density of the foam glass decreases with the progressive increase of sodium silicate solution until it reached a minimum of 0.25 g/cm^3 (90% porosity) at a composition of 12 wt. % sodium silicate solution, after which the density starts to increase once again (fig. 6.28). The compositions of 6 and 32 wt. % have approximately the same density (0.38–0.35g/cm^3). Increasing the amount of foaming agent resulted in decreasing the density of the foam glass until it reached a critical level. Beyond this point the density did not decrease continuously in proportion to increasing foaming agent amount. This is caused by the coalescence phenomenon, where increasing the amount of foaming agent results in extensive foaming. The small pores tend to dissolve into

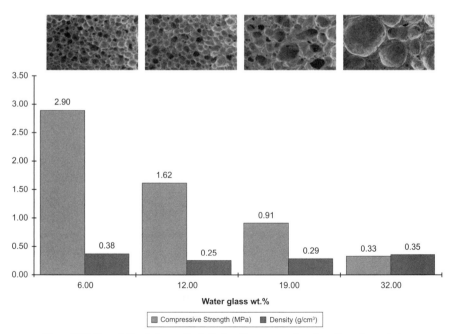

Figure 6.28. The relation between the bulk density, compressive strength, and morphology of foam glass as a function of the amount of sodium silicate solution

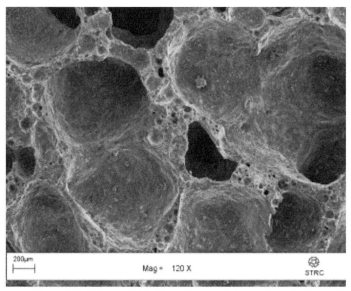

Figure 6.29. Higher magnification of the morphology of foam glass with 19 wt. % sodium silicate solution amount

larger ones in order to decrease the surface energy of the whole system, resulting in coarsening the cellular structure; that is, the pores are larger in size, smaller in number, and the thickness of the cell walls increases. As a result, the density of the foam increases; however, the compressive strength decreases continuously the higher the sodium silicate solution amount. The greater thickness of the cell walls increases the probability of critical flaws, making it weaker than thinner walls. The coarse structure with large pores is a weak structure. This phenomenon is confirmed by the compressive stress curve, which shows that the progressive increase of the sodium silicate solution amount causes a continuous decrease of the compressive strength of the foam glass samples. Figure 6.29 shows the critical flaws that fill the strut.

Effect of soaking time
The effect of soaking time on the foam glass was studied by sintering compositions at 10, 20, 30, and 40 minutes, using white glass with particle size 75µm, sintering temperature 850°C, and 12 wt. % sodium silicate solution. Figure 6.30 shows the variation of the bulk density with soaking time.

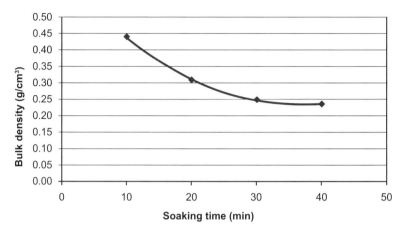

Figure 6.30. Variation of bulk density with soaking time

The results show that the soaking time has a significant effect on the properties of the foam glass. As the soaking time increased, the bulk density of the foam glass decreased. Sintering foam glass for 10 minutes resulted in a dense structure with $\varrho_b = 0.44 g/cm^3$. Increasing the sintering temperature to 20 minutes, then to 30, dropped the bulk density to 0.31, then to 0.25 g/cm^3 respectively. However, further increasing the soaking time to 40 minutes did not significantly decrease the bulk density: from 30 to 40 minutes, bulk density dropped from 0.25 to 0.24 g/cm^3 with the same percentage of porosity (90%).

Figure 6.31 shows that the compressive strength of the foam glass sintered at different soaking times was high. It ranged around 1.6–3.13 MPa. This is caused by the homogeneity of the morphology and the lack of critical flaws in the cell walls. The compressive strength for the foam glass sintered at 10 and 20 minutes was approximately the same (3.13 and 3 MPa respectively). Increasing the soaking time from 20 to 30 minutes caused a significant drop in the compressive strength (from 3 to 1.62 MPa). However, the resistance of the foam glass did not decrease much (from 1.62 to 1.6 MPa) when soaking time was increased from 30 to 40 minutes. Overall, increasing the soaking time from 30 minutes to 40 minutes does not have a significant effect on the foam glass properties. Soaking foam glass for beyond 30 minutes will consume more energy and increase production cost without adding any significant improvement in the properties of the foam produced.

Comparing the effect of sintering temperature vs. soaking time on the morphology of the foam glass (fig. 6.32) shows that the sintering temperature has a more significant effect on the pore morphology. Increasing the sintering temperature from 800 to 900°C increases the size of the pores tremendously because the viscosity of the glass melt decreases. Higher temperature also increases the coalescence phenomenon that coarsens the pore structure. All these factors decrease the strength of the foam glass. On the other hand, changing soaking time does not cause dramatic increase in pore size. It only increases the homogeneity of the pore size distribution, which is why increasing soaking time does not significantly decrease the mechanical resistance of the foam. However, beyond a certain value, further increasing soaking time does not have a significant effect on foam characteristics.

Reducing the sintering temperature to the optimal value is the most important parameter, because it produces foam glass with good characteristics and reduces energy consumption and production costs. Reducing the soaking time to the appropriate value is also an important parameter because it saves energy, in addition to reducing the production cost by reducing the length of the processing cycles. It is crucial to stop the manufacturing process once the optimum characteristics are achieved, in order to interrupt the morphology evolution and to save energy and costs. The experimental results showed that a sintering temperature higher than 850°C and soaking time beyond 30 minutes do not significantly improve foam glass properties. The optimum soaking time is 30 minutes.

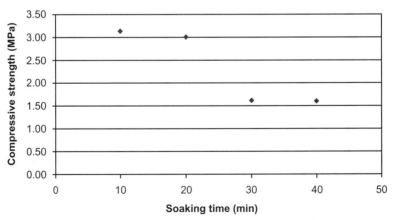

Figure 6.31. Variation of compressive strength with soaking time

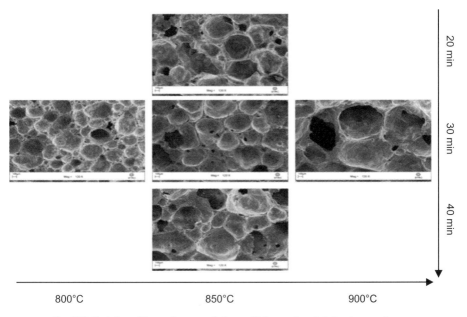

20 min

30 min

40 min

800°C   850°C   900°C

Fig. 6.32. Evolution of foam glass morphology with increasing sintering temperature

Effect of particle size

The effect of particle size on the properties of foam glass was studied by preparing compositions from glass powder with three particle sizes (size 1 = 75µm, size 2 = 150µm, size 3 = 250µm), sintered at three temperatures (850°C, 900°C, 920°C), using white glass with 12 wt. % sodium silicate solution and a soaking time of 30 minutes. Sintering temperatures beyond 850°C were used to allow larger particle sizes to be formed.

Figure 6.33 shows the bulk density as a function of powder particle size at different sintering temperatures. The results show that regardless of the sintering temperature, the bulk density of the foam glass increased with increasing particle size, although the sensitivity of that increase was reduced at the highest sintering temperature of 920°C. The compositions prepared from particle size 1 had low density throughout the different sintering temperatures (0.25, 0.22, 0.24g/cm^3 respectively). The density decreased from 850°C to 900°C, then increased once more at 920°C. This is because at 920°C the structure of the foam is tremendously coarse with thick struts (fig. 6.34c).

The lower viscosity of the glass melt at 920°C along with the coalescence of pores resulted in an increase in pore size. The small size of the particles also caused the pore size to increase tremendously. In general, the sintering rate is strongly affected by the powder particle size. The smaller the particle size, the higher the sintering rate. The finer particles sintered earlier than the larger particles. The density of compositions with particle size 2 and 3 dropped considerably throughout the increasing sintering temperature (size 2: 0.61, 0.40, 0.33 g/cm³; size 3: 1.02, 0.73, 0.52g/cm³ respectively). The higher temperature decreased the density for particle sizes 2 and 3 more than it did for particle size 1. Because particle size 1 is finer, it reached lower viscosity earlier, which is why its density decreased, then increased once again when the structure coarsened. By contrast, sizes 2 and 3 continued to decrease in viscosity and increase in pore size as the temperature increased.

It appears from the morphology (fig. 6.34) that, at the same sintering temperature, as the particle size increases, the pore size distribution becomes less homogeneous and the size of the pore decreases (except in the case of decreasing density for size 1 from 850 to 900°C, as seen in fig. 6.34b). At 850°C, size 1 had the largest pore

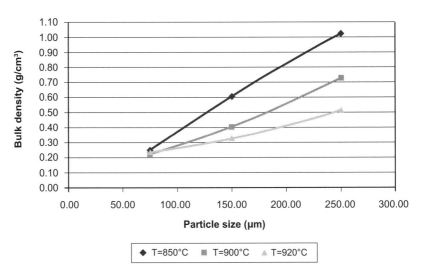

Figure 6.33. Bulk density as a function of powder particle size at different sintering temperatures

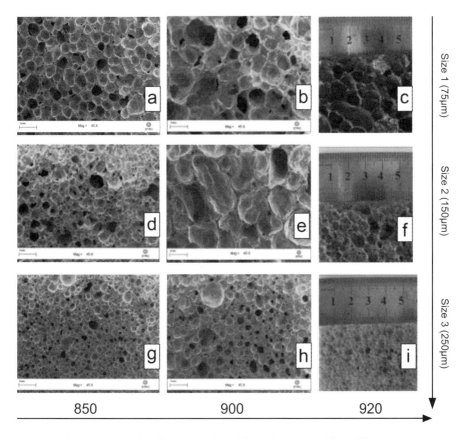

Figure 6.34. Evolution of the morphology of foam glass prepared from different particle sizes and at different sintering temperatures

size but the most homogenous structure (fig. 6.34a, d, g). Size 3 had a dense structure, in which the ratio between the area occupied by the cell struts and the total area was high. It had a relative density of 0.41, which is very high for a foam (fig. 6.34g). Gibson and Ashby (1999) state that beyond a relative density of 0.3, a material is transformed from a foam to a solid that contains isolated pores. Even at 900°C, size 3 still appears as a dense structure with small pore size and inhomogeneous distribution of pore size (fig. 6.34h). The high homogeneity in pore size distribution associated with finer particle size was probably due to the fact that when larger particles were mixed with the foaming agent, the latter agglomerated inside the mixture because of the large voids caused by the large particles, even though the liquid foaming agent was supposed to reduce that effect in comparison with powder foaming agent (Brusatin, Bernardo, and Scarinci, 2004). In general, it appears from the morphology that as the particle size increases, the homogeneity of the foam structure decreases even with increasing temperature.

As the particle size increases, the compressive resistance of the foam increases, except in the two cases that correspond to size 2, whose compressive strength dropped at 900°C and 920°C. The foam glass prepared from size 1 had a relatively good compressive strength of 1.62 MPa at 850°C. Further increasing sintering temperature decreased its resistance until it drastically decreased to 0.34 MPa at 920°C. Although the bulk densities of the foam prepared from size 1 sintered at 850°C and 920°C are approximately equal (0.25, 0.24g/cm³), the one sintered at 920°C is much weaker (fig. 6.35). This is caused by the inhomogeneous structure of the foam sintered at 920°C compared to the foam sintered at 850°C (as shown by the morphology in fig. 6.34a and c). In addition, the foam sintered at 920°C had much coarser structure, with strength-decreasing large pores.

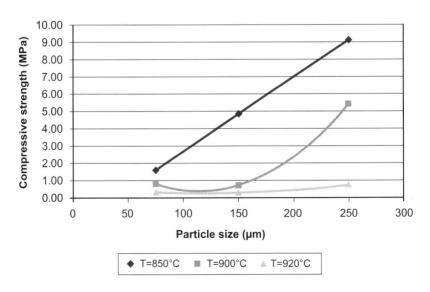

Figure 6.35. Compressive strength of foam glass as a function of glass powder particle size at different sintering temperatures

The foam prepared from size 2 had a dense structure at 850°C (fig. 6.34d), which is why its compressive strength is relatively high (4.87 MPa). However, at higher sintering temperature, its compressive strength dropped severely, to 0.74 MPa at 900°C and then to 0.32 MPa at 920°C. This was expected because the structure of the foam is highly non-homogeneous with large pores and strength-decreasing small pores located around and inside the struts (fig 6.34b). The foam prepared from size 3 was dense whether sintered at 850°C or 900°C; it had high resistance (9.13, 5.42 MPa respectively). However, its strength decreased sharply at 920°C (0.75 MPa) because the structure is inhomogeneous (fig. 6.34i). All the compositions lost their mechanical resistance at 920°C, because at such a high temperature the viscosity of the glass melt was low. The coalescence of pores was amplified and the pore size was large. Consequently, the mechanical strength of all the foam sintered at 920°C was low.

As the particle size increases, higher sintering temperature is required to transform the glass powder into foam. But at higher temperature, the large particle size results in a weak inhomogeneous foam structure full of defects and critical flaws. Although grinding glass powder to finer size will use more energy and cost, the energy consumed to decrease the particle size will be much less than the energy used to increase the sintering temperature. In addition, it will produce relatively stronger foam glass with a homogeneous structure.

Effect of using glass powder of different colors
Glass of different colors has different composition, which might affect the properties of the foam glass produced from them. The effect of container glass color on the properties of the produced foam glass was investigated by preparing samples using green and brown container glass, and comparing them with those prepared using white container glass.

Figure 6.36 shows the variation of the bulk density, compressive strength, and specific compressive strength with the color of the glass (using particle size = 75μm, 12 wt. % sodium silicate solution, sintering temperature 850°C, soaking time 30 minutes).

The specific compressive strength is the ratio between the compressive strength and the bulk density. It describes the strength of the material with respect to its density. This property was calculated for the foam glass in the research study (Liaudis et al., 2009).

The properties of the green and brown glass samples are approximately the same, but they differ significantly from the white glass

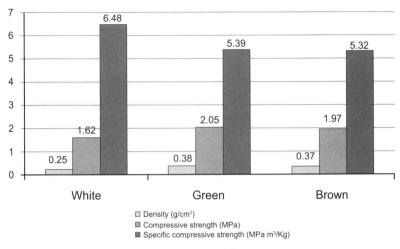

The specific compressive strength values $* 10^{-3}$

Figure 6.36. Relation between bulk density, compressive strength, and specific compressive strength with respect to the color of the container glass

samples. The white glass samples had the lowest density and compressive strength (0.25g/cm^3, 1.62 MPa respectively), as well as the highest specific compressive strength; that is, it is strong with respect to its low density. The green and brown glass samples had bulk densities in the same order of magnitude (0.38, 0.37 MPa respectively) and also similar compressive strength (2.05, 1.97 MPa).

Questions

1. Explain the difference between the solidification behavior of crystalline material vs. non-crystalline material, using glass as an example of a non-crystalline material.
2. Study the growth rate of an engineering product, where glass is a key component, over a period of five years and predict the amount of generated glass waste after its lifetime.
3. Define a cellular material. Explain why the characteristics of foam glass, as a cellular material, make it a highly effective heat-insulating material.

4. List one engineering field where foam glass can be used as an insulator and compare its working properties with other commonly used insulators.

5. Explain, with a schematic representation, the pore formation and growth processes that take place in molten glass during the foaming process.

6. Explain the effect of using different types of foaming agent on the final characteristics of the foam glass.

7. Study the possibility of recycling hazardous glass wastes such as CRTs and fluorescent lamps into foam glass. Analyze whether this will minimize the risk of heavy metal release.

8. Study the effect of material inconsistency factor associated with using recycled glass in producing foam glass and the possibility of using 100% recycled glass in larger-scale mass production.

9. Explain the effect of the shape and size of the pores inside the foam on its thermal and mechanical characteristics.

References

Abdel Alim, D.M. 2009. "Production and Characterization of Foam Glass from Container Glass Waste." MS thesis, Mechanical Engineering Department, American University in Cairo.

American Chamber of Commerce in Shanghai. 2012. "Full Speed Ahead: Alixpartners 2011 China Automotive Outlook." www.amcham-shanghai.org/ftpuploadfiles/publications/autoindustryoutlook/China_Automotive_Outlook_Report_2011.pdf

Bernardo, E., R. Cedro, M. Florean, and S. Hreglich. 2007. "Reutilization and Stabilization of Wastes by the Production of Glass Foams." *Ceramics International* 33, no. 6: 963–68.

Brusatin, G., E. Bernardo, and G. Scarinci. 2004. "Production of Foam Glass from Glass Waste." Proceedings of the International Conference on Sustainable Waste Management and Recycling: Glass Waste, 67–82.

Callister, W.D. 2000. *Materials Science and Engineering: An Introduction.* New York: John Wiley & Sons.

"China Glass Industry Market Report 2011." 2012. Italian Trade Commission. www.ice.it/paesi/asia/cina/upload/174/CHINA%20GLASS%20INDUSTRY%20MARKET%20REPORT%202011.pdf

"CRT Glass to CRT Glass Recycling." 2012. Monitor of Electronics Recycling Issues. www.epa.gov/osw/partnerships/wastewise/pubs/g2gfinal.pdf

Eidukyavichus, K.K., V.R. Matseikene, V.V. Balkyavichus, A.A. Shpokauskas, A.A. Laukaitris, and L.Y. Kunskaite. 2004. "Use of Cullet of Different Chemical Compositions in Foam Glass Production." *Glass and Ceramics* 61, nos. 3–4.

Fernandes, H.R., D.U. Tulyaganov, and J.M.F. Ferreira. 2009. "Preparation and Characterization of Foams from Sheet Glass and Fly Ash Using Carbonates as Foaming Agents." *Ceramics International* 35, no. 1: 229–35.

"Fluorescent Light Bulbs." 2012. Mercury: Consumer and Commercial Products. www.epa.gov/mercury/consumer.htm#flu

"Foamed Glass Technology." N.d. www.enco.ch/glass.htm

"FOAMGLAS Cellular Glass Insulation for Building and Industry." N.d. Foamglas. www.foamglasinsulation.com/

Gibson, L.J., and M.F. Ashby. 1999. *Cellular Solids: Structure and Properties*. Cambridge, UK: Cambridge University Press.

Glass on the Web. 2012. "Review and Prospect of the Glass Industry in China." www.glassonweb.com/articles/article/360/

Glass Recycling UK. 2007. "How It Is Done." www.glassrecycle.co.uk/2007

"Industrial Ecology and Management of Natural Resources." 2007. Class materials for ENGR 592, American University in Cairo, El Gouna trip.

Liaudis, A.S., M.J. Orts Tari, F.J. Garcia Ten, E. Bernardo, and P. Colombo. 2009. "Foaming of Flat Glass Cullet Using Si_3N_4 and MnO_2 Powders." *Ceramics International* 35: 1953–59.

Lund, H.F. 1993. *The McGraw-Hill Recycling Handbook*. New York: McGraw-Hill.

Manevich, V.E., and K.Y. Subbotin. 2008. "Foam Glass and Problems of Energy Conservation." *Glass and Ceramics* 65, nos. 3–4: 3–6.

Mear, F., P. Yot, M. Cambon, R. Caplain, and M. Ribes. 2006. "Characterization of Porous Glasses Prepared from Cathode Ray Tube (CRT)." *Powder Technology* 162, no. 1: 59–63.

Mear, F., P. Yot, R. Viennois, and M. Ribes. 2007. "Mechanical Behaviour and Thermal and Electrical Properties of Foam Glass." *Ceramics International* 33, no. 4: 543–50.

"Miracle Sol Association." 2012. www.miracle-sol.gr.jp/en/en_index.htm

"Municipal Solid Waste." 2012. US Environmental Protection Agency. www.epa.gov/epaoswer/non-hw/muncpl/glass.htm

"Municipal Solid Waste Generation, Recycling, and Disposal in the United States: Facts and Figures for 2010." 2011. www.epa.gov/osw/nonhaz/municipal/pubs/msw_2010_rev_factsheet.pdf

National Center for Resource Recovery, Inc. 1974. *Resource Recovery from Municipal Solid Waste*. Lexington, MA: Lexington Books.

Packaging News. 2012. "UK Leads Euro Growth in Glass Production: Download."
www.packagingnews.co.uk/news/uk-and-turkey-lead-euro-growth-in-glass-production-download/

"Preparation and Placement of Glassphalt." 2012. CWC Best Practices in Glass Recycling. www.p2pays.org/ref/13/12405.pdf

"Region 7: Solid Waste Program." 2012. US Environmental Protection Agency. www.epa.gov/region7/waste/solidwaste/glass.htm

Reindl, J. 2003. "Reuse/Recycling of Glass Cullet for Non-container Uses." Dane County Department of Public Works. www.epa.gov/waste/conserve/rrr/greenscapes/pubs/glass.pdf

Seiner, A.C. 2006. "Foam Glass Production from Vitrified Municipal Waste Fly Ashes." PhD diss. Eindhoven, The Netherlands: Eindhoven University Press. alexandria.tue.nl/extra2/200611231.pdf

Shutov, A.I., L.I. Yashurkaeva, S.V. Alekseev, and T.V. Yashurkaev. 2007. "Study of the Structure of Foam Glass with Different Characteristics." *Glass and Ceramics* 64, nos. 9–10: 3–4.

Smith, W.F. 1996. *Principles of Materials Science and Engineering*. New York: McGraw-Hill.

Solid Waste Management Strategy for Dakahleya Governorate. 1999. "Wastes Characteristics." www.eeaa.gov.eg/seam/Manuals/DakahSolidWaste/Chapter2.pdf

Spiridonov, Y.A., and L.A. Orlova. 2003. "Problems of Foam Glass Production." *Glass and Ceramics* 60, nos. 9–10: 313–14.

Viridor. 2012. "Bottle and Container Glass." www.viridor.co.uk/bottle-and-container-glass

Waste Online. 2012. "Glass Recycling Information Sheet." www.wasteonline.org.uk/resources/InformationSheets/Glass.htm

Williams, P.T. 1998. *Waste Treatment and Disposal*. Chichester, UK: John Wiley & Sons.

Wu, J.P., A.R. Boccaccini, P.D. Lee, M.J. Kershaw, and R.D. Rawlings. 2006. "Glass Ceramic Foams from Coal Ash and Waste Glass: Production and Characterization." *Advances in Applied Ceramics* 105, no. 1.

CHAPTER 7

Innovation in the Plastic Industry 3: Upcycle of Plastic Waste: Natural Fiber Plastic Composites

Mokhtar Kamel and Salah M. El-Haggar

Introduction

Natural and synthetic fibers have played a very important role in the production cost and mechanical properties of different products since plastic technology was first introduced. During the last decade, natural fiber plastic composites (NFPCs) have started to substitute for natural wood in North America, Europe, and China. In addition, they have become an important topic of research because of their properties and advantages. Natural fibers, such as palm, wood, rice, jute, bamboo, kenaf, flax, hemp, cotton, kraft pulp, coconut husk, and many others, are widely used as reinforcements to thermoplastics because they offer a number of advantages over the use of synthetic fibers or the adoption of thermoplastics on their own.

Wang, Sain, and Cooper state that NFPCs have gained increasing interest and their application is becoming more widespread due to their desirable properties (2006). Herrera-Franco and Valadez-González point out the adoption of cellulosic fibers in fiber polymer–reinforced composites, such as coconut fiber, jute, palm, wood, bamboo, and many others. These fibers could be in the natural or the recycled state. The researchers mention their appealing physical and mechanical properties, mainly high stiffness, strength, biodegradability, availability, and cheap cost per unit volume basis. In addition, they are unlike traditional engineering fibers, such as glass and carbon fibers, as they have lower density, higher flexibility, less machine wear, and no health hazards

(Herrera-Franco and Valadez-González, 2004). Lei et al. mention many advantages of using natural fibers as polymer-reinforcing materials, such as high specific strength and modulus, low cost and density, renewable nature, ease of fiber surface modification, and relative non-abrasiveness. They add that natural fiber plastic composites can be obtained from various types of thermoplastics, such as polyethylene (PE), polypropylene (PP), polyvinyl chloride (PVC), polystyrene (PS), and polylactic acid (PLA), compounded with natural fibers, such as wood, kenaf, flax, hemp, cotton, kraft pulp, coconut husk, areca fruit, pineapple leaf, oil palm, sisal, jute, henequen leaf, ovine leather, banana, abaca, and straw (Lei et al., 2007). Sgriccia, Hawley, and Misra suggest the use of natural fibers as a replacement for glass or other traditional reinforcement materials in composites. They consider these materials an attractive alternative to traditional materials because of their superior mechanical properties such as stiffness, impact resistance, flexibility, and modulus. In addition, they are widely available, cause less skin and respiratory irritation, and offer vibration damping and enhanced energy recovery (Sgriccia, Hawley, and Misra, 2008). Jayaraman and Halliwell add that wood fiber waste plastic composites offer attractive mechanical properties and are more cost-effective (Jayaraman and Halliwell, 2009). Alves et al. mention some of the advantages of the use of natural vegetable fibers instead of synthetic fibers. Natural fibers are biodegradable and non-abrasive to processing equipment. In addition, they are lightweight, have high specific strength, are CO_2 neutral, and can be used as acoustic and thermal insulators. In addition, the use of natural fibers in developing countries has offered several social, environmental, and economic advantages, especially considering that the economies of most developing countries are based on agriculture (Alves et al., 2010).

Applications

There are many applications for NFPCs. Lei et al. state that NFPCs have shown high performance in load-bearing applications (Lei et al., 2007). Wang, Sain, and Cooper mention that the main application of natural fiber composites is their adoption in construction materials, such as decking and railing products (Wang, Sain, and Cooper, 2006). Sgriccia, Hawley, and Misra report the interest of US auto makers in using kenaf, flax, and industrial hemp fibers as reinforcements for polymer composites in their products (Sgriccia, Hawley, and Misra, 2008). Alves et al. state that the interest of the automotive industry in using natural fibers as reinforcement instead of glass fibers has increased noticeably in the last decade. They add that automotive interior applications, such as door panels and trunk liners, are the most common ones for reinforced composites. They present a comprehensive study of the advantages of applying jute fiber composites in buggy enclosures (Alves et al., 2010). Jayaraman and Halliwell report that wood-fiber-waste/plastic composites are able to form complex shapes using thermoforming with acceptable mechanical properties (Jayaraman and Halliwell, 2009).

One of the most increasingly used NFPCs is wood–plastic composites (WPCs). WPC applications in various sectors have grown noticeably. Their properties, coupled with increasing environmental awareness and regulations, have led manufacturers to replace conventional woods with WPCs in an increasing number of applications (Adhikary, Pang, and Staiger, 2008). A WPC can be produced in any mold with a variety of shapes and angles, to suit any design (Takatani, Ikeda, and Sakamoto, 2007). Moreover, it can be handled in the same manner as conventional wood, using the same cutting and sawing equipment (Winandy, Stark, and Clemons, 2004). Therefore, any conventional wood workshop can be used with WPC products that have the same functionality as conventional wood (Wechsler and Hiziroglu, 2009). Various WPC products are available in the US market as replacements for certain conventional wood products such as outdoor deck floors, railings, fences, landscaping timbers, siding, park benches, molding and trim, window and door frames, panels, and indoor furniture (Winandy, Stark, and Clemons, 2004).

Wood–plastic composites can also substitute for plain plastics in applications requiring increased stiffness, because the wood fiber elasticity is almost 40 times higher than that of polyethylene and its overall strength is approximately 20 times greater (Bengtsson and Oksman, 2006). It also has higher thermal and creep performance compared to plastics and thus could be used in many structural building applications (Wechsler and Hiziroglu, 2009). Soury et al. (2009) propose the large-scale use of WPC to produce pallets, for which the demand is huge: 400 million wooden pallets a year, accounting for about 86% of all pallets sold worldwide. They point out that, because of the disadvantages of wood-product degradation due to environmental factors, an alternative WPC could be the best option (Soury et al., 2009). WPC began to be utilized in the siding market in 2003, based on studies done in 2002 that revealed that wood occupied about 17% of the materials share of the US siding market, or 960 million square meters (Winandy, Stark, and Clemons, 2004). This opened a potential market for WPC products, which gave a promising performance, similar to wood, over other materials such as aluminum and vinyl (Winand, Stark, and Clemons, 2004).

Material

Wood and plastics (virgin or recycled) of various types, grades, sizes, and conditions are the main materials utilized in WPC production. WPC is composed mainly from a plastic matrix reinforced with wood; other additives are sometimes included. The literature mentions several ingredients of WPC: Najafi, Tajvidi, and Hamadina (2007) state that WPC is a composite made of a natural fiber/filler (such as kenaf fiber, wood flour, hemp, sisal, and so on) which is mixed with a thermoplastic. They add that virgin thermoplastic materials (for example, high and low density polyethylene [LDPE and HDPE], polypropylene [PP], or polyvinyl chloride [PVC]) are commonly utilized. In addition, any recycled plastic that can melt and be processed at a temperature less than the degradation temperature of the wood filler (200°C) could be used to produce WPCs (Najafi, Tajvidi, and Hamadina, 2007). Morton and Rossi (2003) state that the huge majority of WPC utilizes polyethylene. They classify the types of plastic used in WPCs as follow: polyethylene (83%), polyvinyl chloride (9%), polypropylene (7%), and others (1%). Wood flour is obtained from high-quality wasted wood of wood processors; it should be free of bark, dirt, and other foreign matter. Wood species are mainly selected based on regional availability of high-quality flour, and by color. Pine, oak, and maple are the most commonly used in the United States (Clemons and Caufield, 2005). Adhikary, Pang, and Staiger (2008) used recycled and virgin high-density polyethylene (HDPE) with wood flour *(Pinus radiata)* as filler. The HDPE that they used was obtained from a plastics recycling plant, and sawdust was collected from a local sawmill. The fact that WPC ingredients are composed mainly of wood and plastic has led to rapid worldwide growth in its production, due to the huge available quantities of non-utilized plastic and wood wastes.

Plastic waste

The market potential for plastic waste in other utilizations is also large, as plastics constitute the largest share of global municipal and industrial solid waste. According to Kikuchi, Kukacka, and Raschman (2008), most of the plastic waste (78%) goes to landfill; only 22% is recovered. In the United States, the amount of plastic waste in 2005 was calculated as 11.8% of the 246 million tons of municipal solid waste (MSW) generated (USEPA, 2006). In India, plastic makes up to 9 to 12% of MSW by weight, in addition to other waste that may contain much higher proportions of plastics (Panda, Singh, and Mishra, 2010). The majority of the plastic waste generated is disposed (Kikuchi, Kukacka, and Raschman, 2008). However, the continuous growth of worldwide plastic consumption, its short life cycle compared to other products (roughly 40% of plastics have life cycles shorter than 1 month), and legislation in many countries to minimize landfill and incineration encourage recovery of plastic waste instead of disposal (Kikuchi, Kukacka, and Raschman, 2008; Panda, Singh, and Mishra, 2010). Incineration and landfilling alternatives have been rejected outright by several countries, due to their potential danger to the environment in the form of air or land pollution. Incineration and landfilling also result in not closing the loop of cradle-to-cradle and therefore depleting natural resources. As a consequence, the tendency toward recycling has increased (Jayaraman and Bhattacharyya, 2004). Attempts at plastic recovery in 2004 resulted in the recovery of almost 8.25 million tons (39% of total amount of plastics consumed) in Western Europe, and 35,000 tons (13.48% of total imported virgin plastics) in New Zealand (Adhikary, Pang, and Staiger, 2008). In 2005, the United States recycled around 5.7% of the total plastics generated (USEPA, 2006). On the other hand, some states in the United States, such as Michigan, have a recycling rate close to 100% (Beg and Pickering, 2008). In Brazil, around 15% of all plastics consumed are recycled and returned to industry (Beg and Pickering, 2008). This increasing tendency toward recycling plastic bodes well for WPC production in the future.

Wood waste

Wood waste also makes a significant contribution to the total amount of waste arising from various commercial, industrial, and residential activities: scrap lumber, pallets, sawdust, tree stumps, branches, twigs, wooden crates and pallets, building construction and demolition, furniture manufacturing, and many others. Many countries regard it as a major environmental concern. In the United States, a report released in 1995 by the California Integrated Waste Management Board (CIWMB) revealed severe problems with landfill disposing (CIWMB, 1995). It states that the construction and demolition of buildings, which are mainly wood waste, generate almost 12% of all solid waste in California. The average fee for disposing of a ton of waste in a California landfill is about $30 to $35, but disposing of a ton of wood at a wood processing facility may cost only $10. The amount of wasted wood deposited in landfills in some regions of California amounts to 90% of the total wood waste (CIWMB, 1995). Adhikary, Pang, and Staiger claim that a large amount of wood waste is generated from the wood industry at different stages of processing and disposed mostly in landfills. The hazardous content of the wood waste is substantial and takes time to decompose (Adhikary, Pang, and Staiger, 2008). The Department of Environmental Quality (DEQ) in the state of Oregon (US) reports that the alternative to landfill disposal for wood wastes is burning (DEQ, 2009). Wood burners, which were used at first, discharged large amounts of smoke and ash directly into the atmosphere, polluting the air and the environment. They were eventually outlawed and shut down (DEQ, 2009). Recently, a tremendous shift has been made in the area of wood burning with a view to avoiding environmental hazards. Even better, the use of wood waste in WPC helps to overcome the hazards and costs of disposal and burning (Adhikary, Pang, and Staiger, 2008).

Suitability of recycled materials

The issue of using virgin versus recycled (non-virgin) material in producing WPCs is controversial. There are various opinions in the literature regarding the practicality of usage, mechanical properties, physical properties, and even the appearance of the final product. Various comparisons between virgin and recycled materials under many conditions show the pros and cons of recycled material. However, studies based on recycled products are very limited (Adhikary, Pang, and Staiger, 2008) and almost all producers of commercial-scale WPCs are using virgin materials (Klyosov, 2007). This tendency could be due to the fear of obtaining a product with non-controllable properties resulting from impurities, as explained by Yeh, Agarwal, and Gupta (2009).

Adhikary, Pang, and Staiger (2008) used post-consumer HDPE, collected from plastic recycling plants, and sawdust, obtained from a local sawmill. Their study shows the feasibility of making composite panels from recycled HDPE using the hot-press molding technique. They add that the product has superior dimensional stability when compared to virgin HDPE, as well as equivalent tensile and flexural properties. On the other hand, Yeh, Agarwal, and Gupta (2009) showed that wood with recycled acrylonitrile butadiene styrene (ABS) resulted in poor and variable mechanical properties as compared to the corresponding virgin ABS. They add that an unpleasant odor is obtained from recycled material, which could be avoided by adding a thin layer of virgin polymer.

Regarding physical properties, Adhikary, Pang, and Staiger (2008) demonstrate that the panels made from recycled HDPE showed very low water absorption and thickness swelling; thus the products were considered stable in a humid environment. In contrast, Najafi, Tajvidi, and Hamadina (2007) tested water absorption and thickness swelling of WPCs made from sawdust and recycled and virgin plastic (HDPE and PP). The test consisted of 2-hour and 24-hour submersion tests. The results showed that recycled WPCs absorbed more water than virgin, PP absorbed more than HDPE, and the mix of recycled HDPE and PP absorbed the most.

Yeh, Agarwal, and Gupta (2009) and Adhikary, Pang, and Staiger (2008) found variable performance of their final products. Adhikary, Pang, and Staiger explain their results by the different grades and colors of waste stream used and the material contaminants. They also found that the impact of these differences in source materials is still not fully understood, which calls for further investigations and opens a new area of research (Adhikary, Pang, and Staiger, 2008). Yeh, Agarwal, and Gupta believe the problem lies in the reuse of polymers obtained from post-consumer sources. They blame the impurities contained within the materials, which led to decreases in their mechanical and thermal behavior. Based on their findings, such impurities negatively affect the product impact strength and ductility, even if only by 1%. Another problem with impurities in polymers is that the cost of their disposal will be greater than that of virgin material (Yeh, Agarwal, and Gupta, 2009). Therefore, it could be concluded that the main problem in the use of recycled materials is their variable performance. The testing reported in this chapter suggests the same conclusion, as the main cause of variable performance was found to be impurities and contaminants. However, the environmental savings from using non-virgin material—availability, adequate physical properties, and almost no cost—should nonetheless stimulate the use of recycled material instead of virgin.

Testing

The mechanical properties of recycled composites were tested by measuring their flexural properties, as these are the most severely tested characteristics of the product under extreme conditions. They are also of major importance in many international codes related to housing and construction requirements, such as the "Acceptance Criteria for Deck Board Span Ratings and Guardrail Systems" and the "Building Officials and Code Administrators International, 1999 (BOCA) National Building Code" (Klyosov, 2007). Flexural testing is one of the mechanical tests performed on products. (Other mechanical tests could include shear, tensile, impact, and creep tests.) It is frequently carried out on relatively flexible materials such as polymers, wood, and composites (Instron, 2010). It measures the material's behavior when subjected to simple beam loading; in fact, for some materials it is known as the transverse beam test. The specimen in this test is supported at both ends and a load is applied with a known feed rate. Maximum fiber stress and strain are calculated and plotted in a stress–strain diagram, yielding the flexural strength and modulus (Klyosov, 2007; Instron, 2010). Flexural strength and modulus in particular are of major importance for the acceptance criteria of the International Code Council–Evaluation Services (ICC-ES), and for other codes associated with specific applications, such as the AASHTO LRFD Code used for plank decks (Klyosov, 2007; Nowak and Eamon, 2008).

Flexural strength, or modulus of rupture (MOR), is defined as the maximum fiber stress that develops in a tested sample just before cracking or breaking (Instron, 2010). The term 'fiber' here has nothing to do with actual fiber; it refers to the material near the sample surface where the maximum strains occur during loading (Klyosov, 2007). Sometimes, when the elasticity of the material is so high that it will not crack, the flexural yield strength is reported instead of the flexural strength (Instron, 2010). Typically, flexural yield strength is reported when a maximum fiber strain of 5% is reached (ASTM D6272, 2002; ASTM D790, 2008). Flexural modulus (MOE) is defined simply as the stress:strain ratio in the flexural deformation. It is calculated, depending on the apparatus used, from the stress-to-deflection curve slope where the curve has no linear region. A line that is drawn from the origin to the specified point is fitted to the curve to determine slope (Instron, 2010). Klyosov defined its calculation method somewhat differently, as a tangent drawn to the steepest initial straight-line portion of the load deflection curve. Stated more simply, it is a load at which the specimen deflects by 1 inch (Klyosov, 2007).

The flexural test was conducted in this work in accordance with ASTM D4761-05. The goal was to obtain the values of the flexural strength (MOR) and the flexural modulus (MOE). The flexural testing is done typically by four methods or types: 3-point loading, 1/3-point loading (4-point load), 1/4-point loading (4-point load), or uniform loading (Klyosov, 2007). In all cases, the specimen is supported on two edges and a load is applied with a known feed rate. The difference between the methods is mainly based on the number of load noses applied (1 or 2) and the distance between these noses—or, in other words, the maximum bending moment (fiber stress). In the case of 3-point loading, the specimen is loaded with one nose in the middle of the specimen support span (the distance between the two support edges), which positions the maximum axial fiber stress directly under the loading nose. In the 4-point loading, the maximum axial fiber stress is uniformly distributed between the loading noses. The 1/3 (see fig. 7.1) and 1/4 (see fig. 7.2) points refer to the fact that each of the two loads is applied at 1/3 or 1/4 of the length of the support span from the respective ends (ASTM D6272, 2002).

The uniform distributed loading is performed using one of the 3-point or 4-point loads. The uniform load is calculated using standard equations (Klyosov, 2007). The method used in this work consisted of 1/3-point loading (4-point load). Referring to the catalog of the testing apparatus used (InstronBluehillLite), the wood and composites are commonly tested using a 4-point loading flexural test (Instron, 2010), because most of the applications are based on loads distributed all over the product. As for the usage of 1/3-point instead of 1/4, some sources, such as the ASTM D7032 and ASTM D6109, recommended using 1/3-point loading instead of 1/4 (Klyosov, 2007). On the other hand, uniformly distributed testing is not used commonly due to technical difficulties; instead, it is calculated using

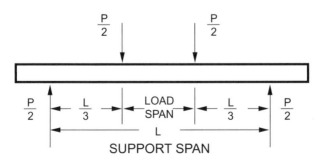

Figure 7.1. 1/3-point loading diagram (4-point load) (ASTM D6272-02, © ASTM International)

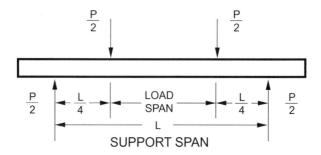

Figure 7.2. 1/4-point loading diagram (4-point load) (ASTM D6272-02, © ASTM International)

standard equations (Klyosov, 2007). It was therefore decided to use the 1/3-point loading (4-point load) because of its easier and more realistic application.

Technologies

Most of the NFPC/WPC technologies mentioned in the literature are very expensive and require high technology such as extrusion and injection-molding techniques (Kamel, 2010). This technology is referred to in this chapter as "first technology." A new, innovative, simple, reliable, and cheap technology to produce NFPC/WPC was developed at the American University in Cairo's Technology and Innovation Lab. This innovative technology is based on compression molding techniques, and is referred to in this chapter as "new technology." It requires fewer production procedures, and less time and cost, than ordinary extrusion and injection-molding techniques.

Experimental Settings

The 1/3 point-loading (4-point load) method was adopted in accordance with the requirements of ASTM D4761-05, which also gave full details for the apparatus and fixtures settings. A graphical display for the test settings (in the standard) made it easy to apply. The experimental settings can be divided into three main parts:

1. *The dimensions of the specimen*: ASTM D4761-05 defines the required span-to-depth ratio as 1:32; however, the standard mentions that this ratio can be changed, but should be documented. The span length chosen for the test was 32cm and the specimen depth was 1cm, where the span is the greatest dimension perpendicular to the direction of

the applied load. The length of the span is determined as the distance between the center lines of edges supporting the specimen (fig. 7.3), while the depth is the smallest dimension parallel to the direction of the applied load and perpendicular to the span (fig. 7.3) (ASTM D4761, 2005). Custom molds were made and used in order to obtain the required dimensions.

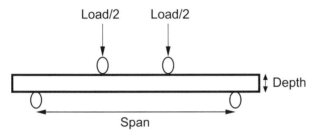

Figure 7.3. Graphical explanation of span and depth

2. *The design of the fixture*: A special fixture was needed, fulfilling the requirements of ASTM D4761, to perform the test. It was custom-made at the AUC workshops. Two equal concentrated points of load application spaced equidistant between the supports were required. The rollers are 3cm in diameter; the distance between their centers is 16cm (fig. 7.4). According to ASTM D4761, the bearing plate should be as wide as the specimen is broad, and its length should be no less than one-half the specimen depth, as shown in figure 7.4. The fixture was then attached to the load ram.

3. *Other requirements:* The ASTM states that the testing machine needs a reaction frame (provided in this case by the supporting edges), a loading mechanism for applying load at specified rate (the ram, with a feed rate of 20mm/min), and a force-measuring apparatus (the computer connected to the machine online). Figures 7.4 and 7.5 depict these components.

Apparatus used: setup, software, and outcome
Instron Model 3300
Instron is a testing machine capable of performing tensile (pull), compression (push), flex (bend), peel, and cyclic types of testing. It is attached to an online computer, through which all the required orders are given (fig. 7.5). The software utilized, InstronBluehillLite, is designed to run Instron's Model 3300 Material Testing Systems.

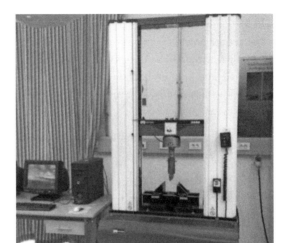

Figure 7.4. Machinery for the flexural test

Figure 7.5. Working station at AUC Testing Lab: the apparatus used (Instron)
and the connected computer

Instron setup

The two edges that form the support span were installed. The custom-made fixture was attached to the load ram and the specimen was positioned between the support edges. The machine was then turned on with the computer. The fixture was adjusted until it touched the specimen surface, and the software settings were adjusted.

InstronBluehillLite settings

The computer settings required four major sets of data regarding the specimen's dimensions, fixture type, feed rate control, and output type, as follows.

1. The specimen's dimensions: the thickness (1cm) and width (10cm) were entered.
2. The fixture type: 4-point fixture type was entered. The support span and loading span (the distance between the centers of loading rollers) were entered, 32cm and 16cm respectively.
3. The feed rate control: a 20mm/min feed rate was entered.
4. The output type: Defined as the measurements (flexural modulus and strength), charts (stress–strain diagrams), and statistics (mean, mode, median, standard deviation) that the researchers wished to obtain.

After the software settings were adjusted, the machine was ordered to start the test. The ram moved downward, applying the load over the specimen via the rollers of the loading fixture. The automatic control and data recordings operated until the completion of the test.

InstronBluehillLite Output

After the completion of the test, the ram returned to its initial position and a new specimen was added. The output of the test consisted of a stress–strain diagram and a table containing the results for each specimen.

Results and Discussion

Table 7.1 represents the results obtained for the mechanical properties, specifically flexural stress and modulus. The test was done for a random sample of 6 parts produced using the new manufacturing technology presented above. These results were compared to the first technology adopted (Kamel, 2010). Table 7.2 presents the mechanical properties of

Table 7.1. Flexural strength and modulus (new technology)

Sample no.	Flexural stress (MPa)	Flexural modulus (MPa)
1	18.1	1440.7
2	19.0	2438.9
3	14.0	1312.7
4	14.3	1517.3
5	14.3	1705.5
6	17.1	2314.3
Range	5.0	1126.2
Standard deviation	2.2	474.8
Average	16.1	1788.2
Median	15.7	1611.4

Table 7.2. Flexural strength and modulus (first technology)

Sample no.	Flexural stress (MPa)	Flexural modulus (MPa)
1	7.1	829.7
2	8.9	878.6
3	6.4	692.7
4	7.8	1826.0
5	4.7	690.2
6	7.3	579.3
7	6.2	513.6
8	8.7	811.4
9	9.4	1118.4
10	10.0	1328.4
Range	5.3	1312.4
Standard deviation	1.63	399.15
Average	7.65	926.83
Median	7.55	820.55

a random sample of 10 parts obtained using the first technique (Kamel, 2010). As noted in table 7.1, the maximum flexural strength and modulus obtained were 19 MPa and 2438.9 MPa. Table 7.2 shows that the maximum flexural strength and modulus obtained were 10 MPa and 1826 MPa, which indicates that the new technology gave higher results at the maximum end of the scale. In comparing the two technologies in terms of central tendency, mean, median, dispersion, range, and standard deviation, it was obvious that the new technology has higher values for central tendencies; however, the dispersion measures were fairly close to each other. This can be explained by the use of different mixtures of ingredients in both technologies; therefore, the relative differences are more or less following a pattern. There is a shift in the central tendency of mechanical properties toward higher values in the new technology, but the relative difference between samples in both technologies is the same. Therefore, it can be concluded that the increase in mechanical properties, in the new technology, is a result of changing the manufacturing processes and has nothing to do with varying the percentages of mixture ingredients.

Summary

Plastic and natural fiber wastes have been a major environmental concern because of their high amount of waste generated, which negatively affects the sustainability of natural resources. One type of use for plastic and natural fiber wastes is wood–plastic composite (WPC), which has been a fast-growing research area in the last decade. WPC products have several environmental and cost benefits. From the narrow environmental point of view, the plastic and wood wastes pose disposal problems, since plastic is non-biodegradable. The recycling of wood and plastic will save energy, reduce pollution and environmental degradation, save trees, and reduce the indirect health costs of hazardous materials. Following the cradle-to-cradle concept, recycling these wastes will close the loop and preserve the environment.

As a cost benefit, these wastes, instead of being treated as worthless by-products, could be an asset for a big industry, generating output with almost no money spent on raw material. In addition to the minimal cost, WPC has a wide range of applications, such as fences, siding, park benches, landscaping timbers, windows and door frames, ponds, indoor furniture, and wooden pallets.

The literature describes several techniques for using WPC, of which the two most important are extrusion and injection molding. The work described in this chapter presents a modified technology based on compression molding. It requires fewer production procedures, and less time and cost, than ordinary extrusion and injection-molding techniques. The output displays better mechanical properties than the products of earlier technology (Kamel, 2010). These results suggest the possibility of even better outcomes in the future.

References

Adhikary, K.B., S. Pang, and M.P. Staiger. 2008. "Dimensional Stability and Mechanical Behaviour of Wood–Plastic Composites Based on Recycled and Virgin High-density Polyethylene (HDPE)." *Composites: Part B* 39: 807–15.

Alves, C., P.M.C. Ferrao, A.J. Silva, L.G. Reis, M. Freitas, L.B. Rodrigues, and D.E. Alves. 2010. "Ecodesign of Automotive Components Making Use of Natural Jute Fiber Composites." *Journal of Cleaner Production* 18: 313–27.

ASTM D790. 2008. *Standard Test Methods for Flexural Properties of Unreinforced and Reinforced Plastics and Electrical Insulating Materials.* Philadelphia: ASTM International.

ASTM D4761. 2005. *Standard Test Methods for Mechanical Properties of Lumber and Wood-Base Structural Material.* Philadelphia: ASTM International.

ASTM D6272. 2002. *Standard Test Method for Flexural Properties of Unreinforced and Reinforced Plastics and Electrical Insulating Materials by Four-Point Bending.* Philadelphia: ASTM International.

Augutis, V., D. Gailius, and G. Balčiūnas. 2004. "Testing System for Composite Wood Based Strips." *Measurement Technologies Laboratory* (Faculty of Telecommunications and Electronics), no. 285.

Beg, M.D.H., and K.L. Pickering. 2008. "Reprocessing of Wood Fibre Reinforced Polypropylene Composites. Part 1: Effects on Physical and Mechanical Properties." *Composites: Part A* 39: 1091–1100.

Bengtsson, M., and K. Oksman. 2006. "Silane Crosslinked Wood Plastic Composites: Processing and Properties." *Composites Science and Technology* 66: 2177–86.

Bledzki, A.K., and O. Faruk. 2004. "Extrusion and Injection Moulded Microcellular Wood Fibre Reinforced Polypropylene Composites." *Cellular Polymers Journal* 4, no. 23: 211–27.

Bouafif, H., A. Koubaa, P. Perré, and A. Cloutier. 2009. "Effects of Fiber Characteristics on the Physical and Mechanical Properties." *Composites: Part A* 40.

CIWMB (California Integrated Waste Management Board). 1995. "CalRe-cycle." Department of Resources Recycling and Recovery. www.calrecycle.ca.gov/

Clemons, C.M., and D.F. Caufield. 2005. "Wood Flour." In *Functional Fillers for Plastics*, edited by M. Xantos, 249–70. Federal Republic of Germany: Forest Products Laboratory. Weinheim: Wiley.

Clemons, C.M., and R.E. Ibach. 2004. "Effects of Processing Method and Mois-ture History on Laboratory Fungal Resistance of Wood–HDPE Composites." *Forest Products Journal* 4, no. 54: 50–57.

DEQ (Department of Environmental Quality, State of Oregon). 2009. blue-book.state.or.us/state/executive/Environmental_Quality.htm

Fabiyi, J.S., A.G. McDonald, M.P. Wolcott, and P.R. Griffiths. 2008. "Wood–Plastic Composites Weathering: Visual Appearance and Chemical Changes." *Polymer Degradation and Stability* 93: 1405–14.

Falk, H. 1997. "Wood Recycling: Opportunities for the Wood Waste Resource." *Forest Products Journal* 6, no. 47: 17–21.

Herrera-Franco, P.J., and A. Valadez-González. 2004. "Mechanical Properties of Continuous Natural Fibre-reinforced Polymer Composites." *Composites: Part A* 35: 339–45.

Instron. 2010. www.instron.us

Jayaraman, K., and D. Bhattacharyya. 2004. "Mechanical Performance of Woodfibre–Waste Plastic Composite Materials." *Resources, Conservation and Recycling Journal* 41, no. 4: 307–19.

Jayaraman, K., and R. Halliwell. 2009. "Harakeke *(Phormium tenax)* Fibre–Waste Plastics Blend Composites Processed." *Composites: Part B* 40: 645–49.

Kamel, M.A. 2010. "Investigating the Physical and Mechanical Properties of Wood–Plastic Composites." MSc thesis, Mechanical Engineering Depart-ment, American University in Cairo.

Kikuchi, R., J. Kukacka, and R. Raschman. 2008. "Grouping of Mixed Waste Plastics according to Chlorine Content." *Separation and Purification Technology* 61, no. 1: 75–81.

Klyosov, A.A. 2007. *Wood–Plastic Composites.* Hoboken: John Wiley & Sons.

Lei, Y., Q. Wu, F. Yao, and Y. Xu. 2007. "Preparation and Properties of Recycled HDPE/Natural Fiber Composites." *Composites: Part A* 38: 1664–74.

McDonough, W., and M. Braungart. 2002. *Cradle to Cradle: Remaking the Way We Make Things.* 1st ed. New York: North Point Press, DuraBook.

Migneault, S., A. Koubaa, F. Erchiqui, and A. Chaala. 2009. "Effects of Pro-cessing Method and Fiber Size on the Structure and Properties." *Composites: Part A* 40: 80–85.

Morton, J., and L. Rossi. 2003. "Current and Emerging Applications for Natural and Wood Fiber Composites." *Seventh International Conference on Woodfiber–Plastic Composites.* Madison, WI: Forest Products Society.

Najafi, S.K., M. Tajvidi, and E. Hamadina. 2007. "Effect of Temperature, Plastic Type and Virginity on the Water Uptake of Sawdust/Plastic Composites." *Holz als Roh- und Werkstoff* 65: 377–82.

Nowak, A.S., and C.D. Eamon. 2008. "Reliability Analysis of Plank Decks." *Journal of Bridge Engineering* 13, no. 5: 540–46.

Panda, A.K., R.K. Singh, and D.K. Mishra. 2010. "Thermolysis of Waste Plastics to Liquid Fuel: A Suitable Method for Plastic Waste Management and Manufacture of Value Added Products—A World Prospective." *Renewable and Sustainable Energy Reviews* 14, no. 1: 233–48.

Sgriccia, N., M.C. Hawley, and M. Misra. 2008. "Characterization of Natural Fiber Surfaces and Natural Fiber Composites." *Composites: Part A* 39: 1632–37.

Soury, E., A.H. Behravesh, E. Rouhani Esfahani, and A. Zolfaghari. 2009. "Design, Optimization and Manufacturing of Wood–Plastic Composite Pallet." *Materials and Design* 30: 4183–91.

Stark, N.M., L.M. Matuana, and C.M. Clemons. 2004. "Effect of Processing Method on Surface and Weathering Characteristics of Wood-flour/HDPE Composites." *Journal of Applied Polymer Science* 93: 1021–30.

Takatani, M., K. Ikeda, and K. Sakamoto. 2007. "Cellulose Esters as Compatibilizers in Wood/Poly(lactic acid) Composite." *Japan Wood Research Society* 54: 54–61.

USEPA (United States Environmental Protection Agency). 2006. *Municipal Solid Waste in the United States: 2005 Facts and Figures.* Official report, Municipal and Industrial Solid Waste Division, US Environmental Protection Agency. Washington, DC: US Environmental Protection Agency.

Wang, W., M. Sain, and P.A. Cooper. 2006. "Study of Moisture Absorption in Natural Fiber Plastic Composites." *Composites Science and Technology* 66: 379–86.

Wechsler, A., and S. Hiziroglu. 2009. "Some of the Properties of Wood–Plastic Composites." *Materials and Design* 30: 4183–91.

———. 2007. "Some of the Properties of Wood–Plastic Composites." *Building and Environment* 42: 2637–44.

Winandy, J.E., N.M. Stark, and C.M. Clemons. 2004. "Considerations in Recycling of Wood–Plastic Composites." RWU4706 Performance Engineered Composites, USDA Forest Service, Forest Products Laboratory, Madison, WI, USA. Fifth Global Wood and Natural Fiber Composites Symposium, Kassel, Germany.

Yeh, S.-K., S. Agarwal, and R.K. Gupta. 2009. "Wood–Plastic Composites Formulated with Virgin and Recycled ABS." *Composites Science and Technology* 69: 2225–30.

Yeh, S.-K., and R.K. Gupta. 2008. "Improved Wood–Plastic Composites through Better Processing." *Composites: Part A* 39: 1694–99.

Youngquist, J., G. Myers, and T. Harten. 1992. *Lignocellulosic–Plastic Composites from Recycled Materials.* Washington, DC: American Chemical Society.

CHAPTER 8

Innovation in the Marble and Granite Industry: Upcycle of Marble and Granite Wastes

Rania Hamza and Salah El-Haggar

Introduction

The dimension-stones industry is one of the oldest and largest industries worldwide. Stone has played a significant role in human endeavors since earliest recorded history and its use has evolved since ancient time. There are two main classes of rocks: calcareous and siliceous. The calcareous material, or marble, includes the whole class of carbonate rocks amenable to sawing and polishing. The siliceous material, or granite, includes the entire set of eruptive rocks with granular structure and poly-mineral composition, irrespective of quartz content (Ciccu et al., 2005).

In terms of composition, marble is a metamorphic rock composed mainly of calcite or dolomite, or a combination of these two carbonate minerals, formed from limestone by heat and pressure in the earth's crust. These forces cause the limestone to change in texture and make-up. This process is called re-crystallization. Granite, a hard natural igneous rock with visible crystalline texture, is formed essentially of quartz and orthoclase or microcline. It is formed from volcanic lava. The principal constituents of granite are feldspar, quartz, and biotite, as described in Minerals Zone (2005).

Since 1990, the world stone industry has expanded at the rate of over 7% per year, due to growth in the building industry and in world population. In 2004, stone production reached 75 million tons (820 million m^2, at a conventional thickness of 2cm), excluding quarry waste.

Granite accounts for less than 40% of overall production, the rest being marble (Ciccu et al., 2005).

The top countries in the production and trade of stone are fast-growing countries including China, India, Turkey, and Brazil, together with Iran and Egypt, along with the traditional leading countries such as Italy, Spain, and Portugal. South Africa and South Korea entered the market in the 1990s. The largest consumers are European countries like Italy, Spain, and Greece, followed by the United States and the United Kingdom (Ciccu et al., 2005).

Nature has endowed Egypt with large deposits of high-quality marble and granite. According to a local survey carried out in 2004, it was estimated that Egypt produces about 3.2 million tons of over 25 different types of marble and granite, placing Egypt among the top eight world producers of raw material. Since 2002, the average annual rate of increase has reached 8.8%. In terms of quarry production and raw export, Egypt is ranked fourth in each, with a share of 4.3% and 6.6% of the total world market of marble and granite, respectively. However, Egypt's share in the international export of total processed material is only 3.7%, and it accounts for only 1.5% of world consumption. Total export from Egypt is approximately 1.5 million tons per year: 0.9 million tons as raw materials and 0.6 million tons as processed products. Egypt is thus the seventh largest exporter in the world, by volume, after China, India, Italy, Spain, Turkey, and Brazil (Ciccu et al., 2005).

The contribution of the natural stone industry to the Egyptian economy has grown tremendously over the past decades and especially since the 1990s. According to Ibrahim Ghali, a dealer for several Italian companies for marble processing and a distributor of marble machinery all over Egypt, there are around 500 large enterprises in this industry and at least 3,000 workshops. According to Medhat Mostafa, president of Sinai International Marble Company in Egypt and vice president of the Association of Marble and Granite producers, capital investments in marble and granite quarries and factories have reached LE 5 billion and more than half a million persons work in this sector as a labor force (Kandil and Selim, 2007).

The regional distribution of factories across Egypt is uneven. According to Ghali, half of all marble producers are concentrated in Cairo. It is often the case that the work is processed in Cairo, but is then transported all over the country, possibly to governorates that lie near to the quarry. This is because the suppliers are located close to the area of demand, which is the head offices of businesses in Cairo. Cairo has thus become the center and main market for marble trading in Egypt. Transportation expenses lead to high overhead costs, due to the lack of adequate regional industry planning. Marble factory owners argue that it is not feasible to situate factories beside quarries, as the quarries are usually located in remote areas, where the lack of water, electricity, and other infrastructure does not permit a minimum living standard for the workers. This is despite the fact that raw marble is easier and less risky to transport than the finished product, which is fragile and has to be handled with care (Kandil and Selim, 2007).

Ciccu et al. (2005) point out that one of the main problems for the marble and granite industry in Egypt is the lack of adequate space for an optimum layout for processing. The existing processing lines are too crowded and ill managed. Infrastructures are poorly developed and often inadequate. It is only recently that the large cluster of Shaq al-Thu'ban near Cairo has been provided with electrical energy from a public network, and it is still waiting for a connection to an industrial water delivery system and for the creation of common waste disposal facilities.

Production Processes

In order to establish an understanding of the waste generated from marble and granite, it will be helpful to review the production process in general. The production of marble passes through several stages. According to Kandil and Selim (2007), the main stages are demonstrated in figure 8.1. The process starts with exploration, followed by extraction, lifting and transportation, arrival at the factory and inventory management, then processing, which includes cutting of the blocks and polishing, and finally cutting into slabs for distribution (Ciccu et al., 2005; Kandil and Selim, 2007).

The processing of marble and granite starts with cutting the raw block into tiles or slabs of varying thickness (usually 2 or 4cm), using diamond blades. To cool the cutting blades and remove the dust created by the cutting process, water is showered on the blades in large quantities. Due to the alkalinity of the wastewater produced, it is not recycled or reused, as it can dim the slabs to be polished. In large factories, where the blocks are cut into slabs, the cooling water is collected in pits, or sedimentation tanks, and stored until the suspended particles settle. The slurry is then removed by trucks and disposed of on the ground. The drying sludge poses an environmental hazard, as it holds large amounts of stone powder that can become airborne after it dries, causing air pollution problems for the surrounding area. Water is also used in the finishing process, when the slabs are finished by either polishing or texturing with the aid of powdered abrasives that scrub the surface of the marble until it becomes smooth and shiny.

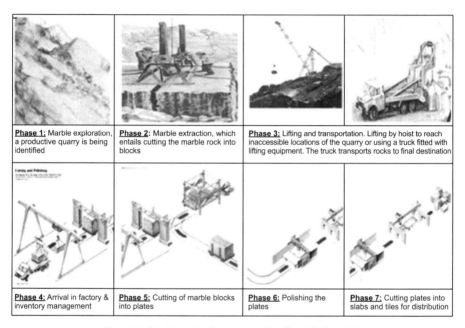

Phase 1: Marble exploration, a productive quarry is being identified	Phase 2: Marble extraction, which entails cutting the marble rock into blocks	Phase 3: Lifting and transportation. Lifting by hoist to reach inaccessible locations of the quarry or using a truck fitted with lifting equipment. The truck transports rocks to final destination	
Phase 4: Arrival in factory & inventory management	Phase 5: Cutting of marble blocks into plates	Phase 6: Polishing the plates	Phase 7: Cutting plates into slabs and tiles for distribution

Figure 8.1. Marble production process (Kandil and Selim, 2007)

Waste Quantification

Actual figures for the quantity of waste produced in Egypt from the marble and granite industry are unavailable, since it is not calculated or monitored by the government or any other party. However, marble waste during block production at quarries is equal to 40–60% (Celik and Sabah, 2008). The total waste generated during the entire process, from mining to manufacturing to the finished product, is in the range of 50–60% of the rock, which is a very high percentage (El-Haggar, 2007).

The largest amount of waste is generated at the cutting stage. When one cubic meter of marble block is cut into 2-cm-thick slabs using a blade 5mm thick, 25% of the marble is converted into powder. The waste generated in the processing stage can be as low as 39% in 300mm x 20mm x free length floor tile production and as high as 53% in 305mm x 305mm x 10mm tile production per 1m³. In other words, as the thickness of the product increases, the proportion of waste is reduced, as described in Celik and Sabah (2008), Monteiro, Pecanha, and Vieira (2004), and Vijayalakshmi, Singh, and Bhatnagar (2003). The waste, in the form of powder, is removed in a mixture of water and other residual materials such as metallic dust and lime, which are used as abrasive and lubricant, respectively. These cakes or sludge have a moisture content around 35–45% (Vijayalakshmi, Singh, and Bhatnagar, 2003; Delgado et al., 2006). Thus, the minimum amount of waste estimated throughout the process of marble and granite production is 40%. The processing waste contributes 20% of the total waste. In an attempt to quantify the total amount of waste generated in Egypt from this industry based on a production of 5.8 million tons (in 2011, based on 8.8% annual growth) and a minimum of 40% waste, one finds that 2.3 million tons of waste is left. In the processing stage alone, the waste is 20%, or 1.2 million tons.

Environmental Impact

Waste generated from the extraction and processing of marble and granite affects the environment in five major areas: topography alteration, land occupation, surface and subterranean water degradation, air pollution, and visual pollution (Celik and Sabah, 2008). Visual pollution and topographical changes are related to quarrying activities. Visual pollution arises from the piling up of fragments generated at the quarrying sites. Topographical changes and disruptions are limited to the area where quarrying operations take place. These areas can be restored, and the impacts are therefore not permanent as long as explosives that produce radioactive materials are not used during quarrying. Besides the environmental problems, excavation and disposal of such large quantities of waste amount to about 25–30% of the total cost of production (Agarwal, 2003).

Although Celik and Sabah (2008) state that marble waste, in general, does not include radioactive by-products, and thus does not induce climate change, it does destroy plant life. This observation is contradicted by Delgado et al. (2006), who believe that marble waste cannot be considered inert (that is, reactive). They base their conclusion upon the conventional leaching tests (DIN 38414 or EN 12457), which confirm that the wastes are alkaline (pH around 12). In addition, the weathering of the worn steel grit and blades used in processing granite releases some quantities of toxic metals like chromium. This endangers the quality of surface and ground waters nearby.

The slurry waste produced from the processing stage usually contains, like the original material, the chemical compounds CaO, MgO, SiO_2, Al_2O_3, Fe_2O_3, Na_2O, TiO_2, and P_2O_5. No global-warming or climate-change gases are released during stonecutting. The dust released is water-scrubbed, and these fine particles can cause pollution unless stored properly in sedimentation tanks and reused later. The dry fine particles, containing white particles of $CaCO_3$, can be easily dispersed under some atmospheric conditions, such as wind and rain, causing air as well as visual pollution. Clay and soils have a high cation exchange capacity and can absorb a high proportion of heavy metals and cations, such as Ca, Mg, K, and Na, and Cl to a lesser extent (Celik and Sabah, 2008). Slurry of particle size less than 80μm is later consolidated as a result of accumulation. A large portion of the particulates remains suspended on the ground surface and does not settle (Celik and Sabah, 2008). The liquid wastes solidify as the water evaporates, but when it rains, the waste leaches into the ground.

The leaching marble slurry can cause water clogging of the soil, increased soil alkalinity, and disruption of photosynthesis and transpiration, leading to a reduction in soil fertility and plant productivity. The internal chemistry of the surviving plants will be altered and their nutritional value poisoned by gases emitted by the industry. This will affect animal as well as human health, as they are interdependent (Celik and Sabah, 2008; El-Haggar, 2007; Vijayalakshmi, Singh, and Bhatnagar, 2003; Pareek, 2007).

In summary, the marble slurry generated during the processing of marble causes the following environmental damage:

- The heaps of slurry remain scattered all around the industrial work area, spoiling the aesthetics of the entire region; the tourism and industrial potential of the area are adversely affected (Vijayalakshmi, Singh, and Bhatnagar, 2003).
- Dry slurry diffuses in the atmosphere, causing air pollution and possibly pollution to nearby water (El-Haggar, 2007).
- The porosity and permeability of the topsoil are reduced tremendously, resulting eventually in waterlogging problems at the surface that prohibit water infiltration. When this happens, the groundwater level drops (Pareek, 2007).

- When the dumped waste dries out, the fine marble dust suspends in the air and is slowly spread out by the wind to the nearby area. It settles on crops and vegetation. It reduces the fertility of the soil by increasing its alkalinity and discouraging vegetation, thus severely threatening the ecology of the area around the marble clusters (Vijayalakshmi, Singh, and Bhatnagar, 2003; Pareek, 2007).
- The high pH of the dry slurry makes it a corrosive material that is harmful to the lungs and may cause eye sores (El-Haggar, 2007; Vijayalakshmi, Singh, and Bhatnagar, 2003). It may also corrode nearby machinery (El-Haggar, 2007).

Marble and granite waste can be utilized in many products as a raw material. The waste is analyzed for its use in added value products such as concrete bricks, composite marble, stone paper, and glass. It can also be incorporated in the cement industry.

Analysis of Marble and Granite Waste
Marble and granite waste material can be classified according to their physical properties, Atterberg limits, grain size, specific gravity, density, water absorption and surface area, and chemical properties.

Atterberg limits
The plasticity of marble and granite particles (powder resulting from slurry) was determined according to ASTM D4318-00 (Liquid Limit and Plastic Limit), and the shrinkage limits according to ASTM D427-98 (Shrinkage Factors of Soils by the Mercury Method). The shrinkage limits (SL) of marble and granite slurry, respectively, as a percentage of dry mass are 23.25 and 27.25, and the shrinkage ratio (R) is 1.51 and 1.47 respectively (Hamza, El-Haggar, and Khedr, 2011a).

Grain size
The grain size of a mixture of marble and granite pieces was determined by sieve analysis according to ASTM C136-01, and by wet sieving (hydrometer) according to ASTM D422-63. The nominal maximum aggregate size is 12.5mm and the coefficient of uniformity (Cu) of the coarse aggregate is 1.9. The fineness modulus (FM) of the mixed coarse sand (A), mixed fine sand (B), and the mixture of 30% A + 70% B, which is selected based on trials for the best gradation curve obtained, are 4.596, 2.755, and 3.307 respectively. Marble and granite slurry powder are of

grain size less than 75 microns, with 90% of the samples of grain size less than 25μm in marble and 35μm in granite; 50% of the particles had a diameter less than 5μm in marble and 8μm in granite. Twenty-five percent of the particles in marble powder, and 20% in granite powder, were less than 2 microns in size, indicating that the samples ranged from clay size to silt, with granite of slightly coarser material than marble. These results show material grain size finer than the finest grain size found in the literature: 90% of the samples are of diameter less than 50 microns, and 50% of the particles had a diameter lower than 7μm, in marble (see figures 8.2, 8.3, 8.4, 8.5) (Hamza, El-Haggar, and Khedr, 2011a).

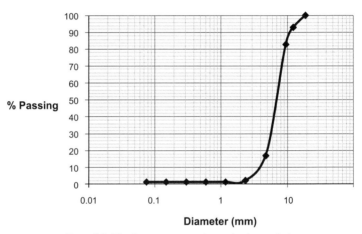

Figure 8.2. Mixed coarse aggregates grain size analysis

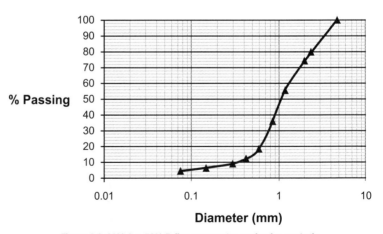

Figure 8.3. 30% A + 70% B fine aggregates grain size analysis

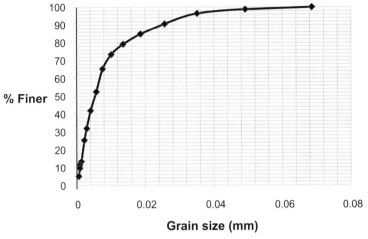

Figure 8.4. Marble slurry powder grain size analysis

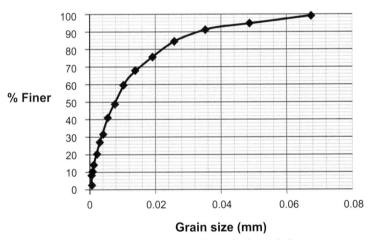

Figure 8.5. Granite slurry powder grain size analysis

Specific gravity, density, and water absorption

The density, relative density (specific gravity), and water absorption of a mixture of marble and granite pieces of diameter greater than 4.75mm were determined according to ASTM C127-07. Specific gravity of pieces with diameters less than 4.75mm was determined according to ASTM C128-07a. Specific gravity of particles with diameters less than 75μm was determined according to ASTM D854-00. The values of specific gravity of the raw material used, both Oven Dry (OD) and Saturated Surface Dry (SSD), are shown in tables 8.1 and 8.2.

Table 8.1. Specific gravity of marble and granite slurry particles

Material	Specific gravity (OD)	Specific gravity (SSD)	% Water absorption
Coarse aggregate	2.407	2.434	1.131
Fine aggregate (A)	2.733 (solids), 2.587	2.632	1.729
Fine aggregate (B)	2.791 (solids), 2.632	2.688	2.145

Table 8.2. Specific gravity of marble and granite pieces

Material	Specific gravity of solids	% Water absorption
Marble slurry	2.768	23.25 (SL)
Granite slurry	2.837	27.24 (SL)

The results are in agreement with the literature (Monteiro, Pecanha, and Vieira, 2004; Vijayalakshmi, Singh, and Bhatnagar, 2003; Almeida, Branco, and Santos, 2007; Saboya, Xavier, and Alexandre, 2007). For slurry powder, the measured specific gravity is as low as 2.55 and as high as 3.0, which is higher than expected for calcite materials. This is due to the presence of abrasive powder (iron grit and lime) used in sawing operations in large units. Thus, the specific gravity of slurry powder varies considerably according to the cutting and processing operations. Aggregate absorption is from 1–2% in coarse and fine stone aggregates, while it is as high as 27% in granite slurry powder. This is due to the high surface area of the particles, which requires high water content for saturation.

Surface area

The surface area of marble and granite particles was determined by the Blaine test according to ASTM C204-07. The measured surface areas of marble and granite slurry particles are 4209 cm^2/g and 4377 cm^2/g respectively, which are comparable to that of cement, 2600–4300 cm^2/g. Although these values are considerably lower than those found in the literature, 0.7–2.5 m^2/g (Corinaldesi, Moriconi, and Naik, 2010), it was indicated that the high surface area of ornamental stone powder should confer more cohesiveness to mortars and concrete.

Chemical analysis

Chemical analysis of slurry powder resulting from processing activities in Shaq al-Thu'ban was determined by INCAx sight by OXFORD instruments, and WD-XRF Spectrometer, PANalytical 2005 (see table 8.3 for marble slurry sample resulting from gang saw, table 8.4 for granite slurry sample resulting from gang saw, and table 8.5 for granite slurry sample resulting from multi-disk operation).

Marble powder shows calcium oxide as the major component (> 49%) with loss of ignition (LOI) around 40%, and small amounts of SiO_2 (< 5%), MgO (< 3%), and Fe_2O_3 (< 2%), as indicated in the literature. On the contrary, granite shows SiO_2 as the major component (> 60%), with a much lower level of LOI (< 2%), with Al_2O_3 values between 6 and 14%, CaO values of 0 to 6%, and traces of Na_2O and K_2O (0 to 3.5%), in agreement with the values indicated in the literature (Vijayalakshmi, Singh, and Bhatnagar, 2003; Rizzo, D'Agostino, and Ercoli, 2008). It is worth mentioning that the chemical analysis of granite slurry resulting

Table 8.3. Chemical analysis of marble gang saw sample

Concentration of major constituents	Weight, %
SiO_2	0.57
Al_2O_3	0.16
Fe_2O_3	0.11
MgO	0.2
CaO	55.26
Na_2O	0.05
SO_3O	0.06
ZrO_2	0.01
P_2O_5	0.02
SrO	0.03
Cl	0.01
LOI	43.52

from cutting operations using gang saws showed higher values of Fe_2O_3 (7.73), compared to 0 to 6.5% in the literature (Monteiro, Pecanha, and Vieira, 2004; Torres et al., 2009; Menezes et al., 2005), which indicate the use of iron grit as abrasive material in the cutting procedure. In addition, some granite samples show small values of CaO, around 3% (see table 8.4), which indicates a mix of marble and granite waste.

Table 8.4. Chemical analysis of granite gang saw sample

Concentration of major constituents	Weight, %
SiO_2	69.88
TiO_2	0.05
Al_2O_3	12.21
Fe_2O_3	7.73
MgO	0.07
CaO	3.17
Na_2O	3
K_2O	3.65
Cr_2O_3	0.07
P_2O_5	0.03
SO_3	0.05
MnO	0.07
Cl	0.01
LOI	—
Trace elements	Ppm
Cu	44
Rb	147
Sr	57
Y	36
Zn	46
Zr	46

Utilization of Marble and Granite Waste in Concrete Bricks

Concrete bricks can be the best application to utilize marble and granite waste in large quantities to replace the conventional sand and aggregates. Normally, aggregates in concrete bricks are dolomite as the coarse aggregate, and sand as the fine component. These can be replaced by marble and granite waste aggregates of different sizes with the addition of slurry powder. The slurry powder, with its very low grain size (less than 70 microns) and its high surface area (more than

Table 8.5. Chemical analysis of granite multi-disk sample

Concentration of major constituents	Weight, %
SiO_2	69.99
TiO_2	0.34
Al_2O_3	14.01
Fe_2O_3	2.98
MgO	0.82
CaO	1.68
Na_2O	3.57
K2O	4.3
P_2O_5	0.1
SO3	0.02
MnO	0.07
Cl	0.02
LOI	1.9
Trace elements	Ppm
Cu	90
Rb	111
Sr	172
Y	58
Zn	64
Co	79
Nb	38

4200 cm^2/g), can add cohesion and micro-filling ability to the bricks (Hamza, El-Haggar, and Khedr, 2011a; 2011b).

Stone waste aggregate replacing sandstone aggregate in concrete

Stone pieces resulting from the cutting process are of good quality stone, with chemical and physical properties that allow it to be used in cement concrete to produce value-added products (Hamza, El-Haggar, and Khedr, 2011a; 2011b).

Studies were carried out by the civil engineering department of Kotah Engineering College in Rajasthan on the utilization of Kotah stone aggregate in cement concrete to replace sandstone aggregate. The mix design was 1:1.5:3 cement to fine aggregate to coarse aggregates. The results showed that Kotah stone waste aggregate can be used to make cement concrete, and that the loss in compressive strength when compared to cement concrete (sandstone) is only 15.7% after 60 days. The average compressive strength of the sandstone mix and the Kotah stone were 28.97 MPa and 24.41 MPa respectively (Agarwal, 2003).

Slurry waste replacing fine aggregate in concrete

The use of marble slurry in concrete works reveals enhanced properties. When 5% of the initial sand content was replaced by stone slurry, a compressive strength 10.3% higher after 7 days, and 7.1% higher after 28 days, were measured, when compared to concrete mixture without slurry addition (Almeida, Branco, and Santos, 2007). This increase is attributed to the micro-filling ability of slurry, which promotes effective packing and greater dispersion of cement particles, and accelerated formation of hydrated compounds, resulting in a significant improvement of compressive strength at earlier ages (7 days). However, at higher percentages (10%, 15%, and 20%) compressive strength is reduced from 3.6% to 10.6% at 7 days of age, and from 6.7% to 8.9% at 28 days of age, when compared to mixture with no slurry. This was because the low water/cement ratio meant that the available space for accommodating hydrated products was insufficient, thus inhibiting chemical reactions. When the percentage of slurry increased above 20%, the compressive strength was considerably reduced, as the micro-filler effect did not prevail and inappropriate grading caused lower results. For 100% substitution of fine aggregates with stone slurry, the results were significantly reduced, to 40.9% less for 28 days and 50.1% less for

7 days. Thus, full substitution of fine aggregate with stone slurry is not reliable when compressive strength is a critical aspect.

The improvement in compressive strength due to the micro-filling effect induced by stone slurry particles also extended to the splitting tensile strength tests, where a 14.3% increase was detected for a mixture with 5% slurry. Higher percentages of stone slurry led to negative results for splitting tensile strength. However, the tests showed that the splitting tensile strength is less sensitive to high content of slurry particles than compressive strength: full substitution resulted in a 28.6% reduction relative to 0% mixture.

With respect to elasticity, test results were in accordance with the other mechanical properties. The 5% mixture showed better behavior (6.2%) than the 0% mixture, and the higher the slurry incorporation, the more prominent the negative effect on elasticity. This was attributed to the lower value of elasticity for cement paste (half) compared to aggregates, as when incorporating the very fine particles of slurry, the paste could be considered as increased, thus reducing the modulus of elasticity of the entire mixture.

On the other hand, the properties of concrete mixtures are further improved at 15% substitution of fine aggregate for marble slurry. Samples with 5%, 10%, and 15% substitution were tested. The 15% sample showed the highest compressive strength at 28, 90, and 360 days (40, 50, and 60 MPa respectively), compared to the control specimens (0% slurry) with compressive strength values of 28, 32, and 35 MPa respectively. It is interesting to note that the reported average compressive strength for samples with 15% slurry was 12% and 20% higher than samples with 5% and 10% slurry respectively at 28 days, 8% higher at 90 days, and 5% and 8% higher at 360 days respectively. It was observed that the relative strength values of all specimens (including the control specimen) were almost equal at early ages, while the relative strengths of marble slurry samples were higher after 28 days.

Marble slurry samples showed an obvious increase in sulfate resistance of concrete. Research studies report that both natural and artificial pozzolans can contribute to increasing chemical resistance (Binici, Kaplan, and Yilmaz, 2007). The compressive strengths at 12 months of sulfate attack for samples with slurry were greater than those obtained with normal aggregate concretes, and increasing the additive (slurry) content caused significant increases in sulfate resistance. The compressive strength of samples after 28 days of exposure to tap water was 35.2

MPa for control specimen, 56.3 MPa for 5% slurry, 58.6 MPa for 10% slurry, and 60.7 for 15% slurry. Values were reduced to 15.1 MPa (58% reduction), 41 MPa (28% reduction), 48.4 MPa (18% reduction), and 52 MPa (15% reduction) after 12 months of sulfate exposure.

The results of tests for abrasion resistance show that specimens with slurry additives exhibit better abrasion resistance (15% slurry enhanced the abrasion ratio to 52%, 45%, and 42% at 28, 90, and 360 days respectively, compared to 100% for 0% slurry at all ages at 60 minutes of abrasion). This can be attribute to the fact that dust replacement allowed a good interfacial condensed matrix. Replacing sand with limestone dust did not enhance the abrasion resistance like the use of marble dust; on the contrary, a 15% limestone dust showed less abrasion resistance than the 10% limestone samples, indicating that both the dust type and the amount of dust have a noticeable effect on abrasion resistance of concrete, and that the grain size distribution of aggregate is an important feature in this process.

With respect to water penetration, marble slurry additive of 15% considerably decreased the depth of penetration, from 13mm in the control specimen to 4mm at 15% slurry. This is attributed to the dense interface between the aggregates and the cement base caused by the adherent ability of the slurry that combines with concrete elements to form more dense and homogenous forms of concrete, giving less porous structures (Binici, Kaplan, and Yilmaz, 2007).

Higher amounts of sand substitution for slurry were tested by Misra and Mathur (2003), as the mechanical properties of concrete were enhanced when up to 40% of the slurry was replaced by fine aggregates: 19% (from 27 to 32.2 MPa) and 49% (40.2 MPa) increase in compressive strength at 7 days for 20% and 40% slurry respectively, and 11% (from 36.3 to 40.3 MPa) and 26% (45.7 MPa) increase at 20% and 40% slurry respectively at 28 days. There were also gains in flexural strength, but much less prominent (< 5%, from 4 to 4.19 MPa). However, the workability of the concrete mix decreased significantly, necessitating the use of a higher w/c ratio to maintain the same slump. The w/c ratio increased from 0.53 to 0.55 when 40% slurry particles were substituted for sand in the mix, maintaining the slump at 25mm. Other mechanical properties related to durability were also improved at 40% slurry substitution for sand: abrasion resistance (0.21% loss at 0% replacement and 0.187% loss at 40% replacement), water soaking, freezing and thawing, sulphate attack, and heating–cooling.

Slurry waste partially replacing cement in concrete

Kotah Engineering College investigated the effect of replacing cement with slurry waste at percentages of 15, 30, and 45. In addition, tests were performed with Kotah stone replacing sandstone in a mix with ratio 1:1.5:3. The compressive strength of the control mixtures of sandstone only, and Kotah stone only, were 28.97 and 24.41 MPa respectively, while 15%, 30%, and 45% slurry replacement caused a drop in compressive strength to 23.9, 27.22, and 10.36 MPa respectively. The results showed that stone slurry can be used in cement concrete with slurry replacement of cement up to 30%, with loss in compressive strength of only 6%, yet a cost saving of over 21% (Agarwal, 2003).

Mix design

The mix design incorporates around 10 wt % cement, 30 wt % fine aggregates with ratio 3:7 A:B (FM 3.307), 50 wt % coarse aggregates, and marble (M) or granite (G) slurry powder of 10, 20, 30 and 40 wt %, with proportional redistribution of coarse and fine aggregates to accommodate the added slurry powder beyond 10%. In addition, 0% slurry brick and control brick, with conventional dolomite coarse aggregates and sand, were tested (see table 8.6).

Sampling and testing of concrete bricks

The bricks produced have the dimensions of 250mm length, 120mm width, and 60mm height, in accordance with the brick dimensions specified by the Egyptian code for masonry works. Three samples of each brick formula were tested after 7 and 28 days for compression, moisture, absorption, and durability (heating and cooling cycles and saturated salt solution, sodium chloride, immersion cycles followed by heating, on a 24 ± 2 hours cycle for 7 days). Results are compared to ASTM C140, the Egyptian code, and the control samples. In addition, brick abrasion resistance is compared to ASTM C902-09.

Results

The results show that the marble and the granite slurry samples yield similar mechanical properties, in terms of compressive strength, and similar physical properties, in terms of density and absorption. In compressive strength, although both marble and granite show similar results, granite slurry samples show slightly higher values, as illustrated in figures 8.6 and 8.7, which is predictable due to the higher strength of

Table 8.6. Mix design for concrete bricks

Mix ID	Cement (kg/m³)	Slurry (kg/m³)	Fine aggregates (kg/m³)	Coarse aggregates (kg/m³)
Control	300	—	800	1000
Zero	235	0	704	1173
M10, G10	232	232	696	1160
M20, G20	220	441	579	965
M30, G30	210	630	472	787
M40, G40	204	818	383	639

natural granite stone and the apparent stronger bond with cement paste. This increase in strength in granite slurry bricks compared to the marble slurry ones is around 10%, 11%, 14%, and 33% in 10%, 20%, 30%, and 40% samples respectively at 7 days. In the 28-day test, the increase is 9%, 23%, 9%, and 48% for 10%, 20%, 30%, and 40% samples respectively. It is worth mentioning, however, that the 40% granite slurry samples (G40) show much higher values of compressive strength (33% and 48%) than marble slurry. This may indicate that granite slurry has a better interface with cement paste in the mix beyond the purely physical micro-filling action. The effect is more noticeable at higher percentages of granite fines. In addition, both marble and granite samples show similar trends in terms of the degree of strength achieved after 7 days when compared to 28 days. For example, the 10% slurry samples, both marble and granite, achieve 80% of the 28-day strength, whereas the 20% marble and granite slurry samples achieve 83% and 72% of the 28-day strength, respectively (Hamza, El-Haggar, and Khedr, 2011a; 2011b).

Comparing compressive strength with the control sample at 28 days, the 10% marble slurry (39.4 MPa) and granite slurry (43.48 MPa) samples yield results close to that of the control (39.6 MPa). The 20% granite slurry samples also show similar results (36.95) to the control. These results emphasize the positive effect of granite slurry on brick samples, which reaches its optimum at 10% slurry incorporation, while at higher percentages, agglomeration of slurry started to appear, acting as media discontinuities and thus decreasing the compressive strength of the samples. Zero slurry samples showed the lowest compressive strength of all samples. This is due to the poor grain size distribution and the lack of filling materials (Hamza, El-Haggar, and Khedr, 2011a; 2011b).

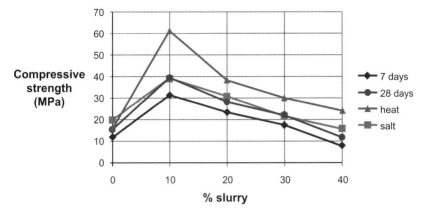

Figure 8.6. Compressive strength of granite slurry samples

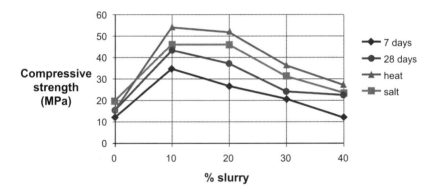

Figure 8.7. Compressive strength of marble slurry samples

All of the samples are acceptable, in terms of compressive strength, according to the Egyptian specifications for structural requirements (7 MPa). However, as compared to ASTM C55, the control (39.6 MPa), M10 (39.4 MPa), M20 (28.3 MPa), G10 (43.5 MPa), G20 (37.0 MPa), and G30 (24.1 MPa) are acceptable for grade N (for architectural veneer and facing units in exterior walls and for use where high strength and resistance to moisture penetration and severe frost action are desired: 24.1 MPa, average of 3, and 20.7 MPa, individual unit). M30 (22.0 MPa) and G40 (22.8 MPa) are acceptable for Grade S (for general use where moderate strength and resistance to frost action and moisture are required: 17.3 MPa, average of 3, and 13.8 MPa, individual unit). M40 is rejected for falling below the limits of Grade S. As for density, most samples, including the control, are of normal weight (> 2000 kg/m^3), according to both the Egyptian specifications and ASTM C55, except for M30, G30, and G40, which are of medium weight (1680–2000 kg/m^3) (Hamza, El-Haggar, and Khedr, 2011a; 2011b).

Both heating and salt solution soaking and heating cycles increased the compressive strength of all samples with different ratios. Thus, it can be concluded that heating and cooling cycles did not adversely affect samples; on the contrary, they enhance compressive strength. This may be attributed to the accelerated cement hydration at higher temperatures, which apparently counters the heat-associated volumetric changes (Hamza, El-Haggar, and Khedr, 2011a; 2011b).

Absorption is the major drawback of incorporating slurry in bricks. Although the Egyptian specifications for concrete bricks do not impose limits for absorption, they specify a maximum of 16% for wall-bearing bricks, and 20% for non-wall-bearing fired-clay bricks. All samples show absorption of less than 15%. With respect to ASTM specifications, zero, M10, G10, M20, and G20 fulfill the requirements for grade S (208 kg/m^3, 10.1% for normal weight, and 240 kg/m^3 for medium weight), with absorption values of 168 kg/m^3, 168 kg/m^3, 185 kg/m^3, 193 kg/m^3, 201 kg/m^3, 179 kg/m^3, and 184 kg/m^3 (see fig. 8.8) (Hamza, El-Haggar, and Khedr, 2011a; 2011b).

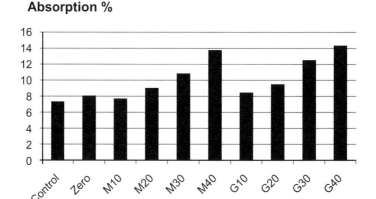

Figure 8.8. Water absorption for marble and granite slurry samples

Regarding the abrasion resistance of bricks for pedestrian and light traffic use according to ASTM C902, the control sample, zero, M10, M20, M30, G10, and G20 are all classified as class MX (brick intended for exterior use where resistance to freezing is not a factor), type II (brick subjected to intermediate abrasion). G30 is classified as class MX, type III (brick subjected to low abrasion), while G40 is classified as class NX (brick not intended for exterior use but which may be acceptable for interior use where protected from freezing when wet), type III (Hamza, El-Haggar, and Khedr, 2011a; 2011b).

Utilization of Marble and Granite Waste in Composite Marble

Agglomerate marble (or composite marble, or cultured marble) is the designation for products that bind pieces of natural marble together with specially formulated polyester resin. This process allows the reconstruction of large recycled marble blocks, similar to the ones extracted from quarries, both in quality and visual aspects, which can be processed as natural stone (Almeida, Branco, and Santos, 2007). Research was conducted to partially substitute calcium carbonate for marble slurry for up to 6% of the total compounds, a procedure which proved to be technically viable (Almeida, Branco, and Santos, 2007). Kumar, Suryanarayana, and Venkatesh (2003) indicate that up to 50% marble slurry powder can be used in a polymer composite marble to produce bathtubs, kitchen sinks, shower trays, and washbasins, with cost savings of up to 30% compared to conventional ones.

Agglomerate marble is an Italian contribution, widely known as compound stone, in which 84–91% is stone aggregate by overall weight. The technology involves compaction by vibro-compression, under vacuum, of a mixture formed from stone aggregate and binding paste. In this system, composite marble blocks of various sizes, such as 308x122x88cm or 186x125x87–88cm, are first manufactured by compaction. These blocks are then sawn into slabs of various thicknesses (9mm, 14mm, 20mm, 30mm, and so on). The slabs are sawn to the desired sizes, honed, chamfered, polished, dried, and cleaned to give finished products. The product can be colored by adding various pigments. The product is claimed to have outstanding geo-mechanical properties such as high compressive strength, low water absorption, mechanical resistance, resistance to chemical agents, heavy density, and so on, which are comparable to natural stone (Agarwal, 2003).

Various experiments have been conducted to examine the properties of composite marble. Borsellino, Calabrese, and Di Bella (2008) investigated the effects of marble powder concentration and type of resin on the performance of marble composite structures. The test studied powder ratios of 60%, 70%, and 80%, and two resin types: epoxy and polyester. The sample tiles were of size 200x100x10mm molded in a wood mold after a heterogeneous mixing of powder/resin and kept in rotation until full cure of the matrix to avoid depositing of marble powder. Static flexural tests, stress–strain tests, were conducted on the specimens. The 60% powder sample with an epoxy resin showed superior properties to the monolithic marble in terms of stress (25 MPa), impact strength (0.27 J/cm^2), percentage of water absorption (0.16%), and stain resistance for various substances: Coca-Cola, coffee, detergent, lemon, oil, wine, butter, ketchup, and mayonnaise (no alteration on the surface after 2 hours of treatment). It can be concluded that the results of the tests are highly dependent upon the resin properties and amount.

Polymer composite marble was also prepared in the Regional Research Laboratory (RRL), Bhopal, in central India. The results conform closely to the results obtained by Borsellino, Calabrese, and Di Bella (2008) except for compressive strength, where the Bhopal results showed much higher values. Polymer composite marble at RRL was tested for density (1.96 g/cm^2), moisture content (0.2–0.4%), modulus of rupture (21–26 MPa), tensile strength (23–25 MPa), compressive strength (77–96 MPa), and water absorption at 2 hours (0.15– 0.40%). Tests also proved that the polymer composite marble is fire retardant, shows no change in dimensions when exposed to boiling water, and is chemically inactive (Vijayalakshmi, Singh, and Bhatnagar, 2003).

Mix design

The composite marble matrix is composed of marble or granite slurry powder (SL), coarse (A) and fine (B) sand, bonded by a polyester resin. The crushed marble and granite particles and pieces are mixed with polyester resin, along with its hardener and accelerator. The percentages of the composite mix are detailed in table 8.7, where composite marble (CM) samples are composed of waste aggregates and slurry bonded with polyester in various percentages (14, 16, 18, and 20 wt. %) to find the optimum value. Composite marble slurry (CMS) samples, composed of slurry powder and resin, are also tested with slurry content

Table 8.7. Composite marble mix design

Sample code		Resin (wt %)	SL (wt %)		Fine sand (wt %)		Coarse sand (wt %)		Total (wt %)	
CM	Poly14	14.89	20	21.28	20	21.28	40	42.55	94	100
	Poly16	16.67	20	20.83	20	20.83	40	41.67	96	100
	Poly18	18.37	20	20.41	20	20.41	40	40.82	98	100
CMS	Poly20	20	20	20	20	20	40	40	100	100
	Poly30	30	70		—		—		100	
	Poly35	35	65		—		—		100	
	Poly40	40	60		—		—		100	

of 60, 65, and 70 wt. %, with 40, 35, and 30% resin respectively. Setting time of samples is 15–25 minutes at room temperature (22–25°C).

Sampling and testing of composite marble

After drying, the samples were removed from molds and sawn, if necessary, into the required dimensions. Physical and mechanical properties were determined according to European standards. Results were compared to ASTM specifications, if applicable, and to commercially available marble and granite properties. Six 50mm cubes were tested for each mix design for real density and porosity of composite marble according to EN 1936:1999, water absorption at atmospheric pressure of composite marble according to EN 13755:200, and compressive strength of composite marble according to EN 1926:1999. Ten specimens of dimensions 150mm x 75mm x 25mm were tested for flexural strength of composite marble under concentrated load according to EN 12372:1999. Six specimens of dimensions 200mm x 200mm x 30mm were tested for each mix design for rupture energy of composite marble according to EN 14758:2004.

Results

Poly16 (16% polyester resin and 84% aggregate and slurry powder) showed the highest density (2126 kg/m^3) and the lowest porosity (0.5%). For composite slurry marble, poly30 (30% polyester and 70% slurry powder) showed the highest density (1860 kg/m^3) and the lowest porosity (0.72%). The lowest water absorption is in the poly16 samples. For composite slurry marble, poly30 showed the lowest water absorption (0.37%). Poly14, with the highest percentage of aggregates, showed the highest compressive strength. At higher slurry percentages, agglomeration of slurry started to decrease the bond that the slurry powder creates between the resin and the aggregate matrix. This caused stress concentration, thus yielding lower compressive strength. This effect becomes vivid when a matrix of slurry powder and resin is considered, as shown in the composite slurry marble samples, where all composite slurry marble failed to resist compression, and behaved like putty. In flexural strength, all samples showed similar values. However, the highest value of flexural strength appeared in poly14 samples (14.8 MPa). For composite slurry marble, poly30 showed the highest flexural strength (5.1 MPa). All samples showed similar values of rupture energy; however, the highest value appears in poly16 samples. Composite marble slurry showed the highest value of rupture energy at poly40 (16.1 joule). The results indicate that poly16 showed the best physical and mechanical properties: the lowest porosity (0.5%), the lowest water absorption (0.24%), the highest density (2126 kg/m³), and the highest rupture energy (13.4 joule). Although it did not show the highest value for compressive strength, at 71.63 MPa, this value is still higher than the required strength for marble in ASTM C503-08a. Results are shown in table 8.8 and illustrated in figures 8.9, 8.10, 8.11, and 8.12.

The composite marble sample exceeds the values required by ASTM C503-08a for compressive and flexural strengths, 52 and 7 MPa respectively. However, the water absorption slightly exceeds the maximum limit of 0.2%. Although the density of composite marble is lower than that required by ASTM C503-08a, 2595 kg/m³, this is an expected finding resulting from the light density binder; it can be considered advantageous in lightweight applications.

For the composite marble slurry samples, all samples reveal unsatisfactory results except for the rupture energy (see table 8.8), which shows the highest values of all composite marble samples. It is not recommended for use in most marble applications due to its substandard compressive strength, low values of flexural strength, and higher porosity and water absorption.

Composite marble values are comparable to commercially available marble and granite. According to test results of the Marble and Granite Technology Center for commercial marble and granite, marble has open porosity, absorption, compressive strength, flexural strength, and rupture energy values of 2.7%, 1.34%, 66 MPa, 5.7 MPa, and 4.5 joule respectively, while granite has values of 0.3%, 0.12%, 136 MPa, 14.6 MPa, and 5.6 joule respectively. Thus, composite marble surpasses commercial marble in porosity (0.5%), absorption (0.24%), compressive strength (72 MPa), flexural strength (13.5 MPa), and rupture energy (13 joule), yet falls behind granite in all properties except rupture energy.

Table 8.8. Composite marble test results

Sample	Apparent density (kg/m³)	Open porosity (%)	Water absorption (%)	Compressive strength (MPa)	Flexural strength (MPa)	Rupture energy (joule)
Poly 14	2049.35	0.7	0.33	79.87	14.8	12.1
Poly 16*	**2126.23**	**0.5**	**0.24**	**71.63**	**13.47**	**13.4**
Poly 18	1806.78	0.8	0.37	71.04	14.41	12.26
Poly 20	2065.73	0.87	0.44	64.38	14.5	12.34
Poly 30	1860.47	0.72	0.37	—	5.11	14.38
Poly 35	1702.52	0.9	0.52	—	4.56	15.69
Poly 40	1603.77	0.75	0.45	—	3.3	16.09

* Best performance

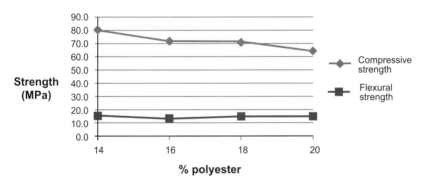

Figure 8.9. Compressive and flexural strength of composite marble at different polyester content

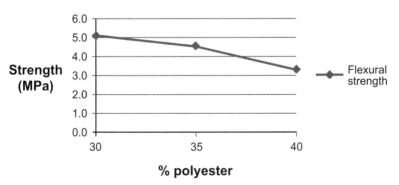

Figure 8.10. Flexural strength of composite slurry marble at different polyester content

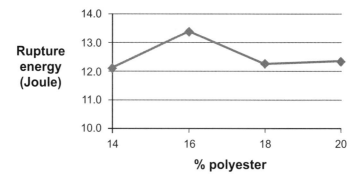

Figure 8.11. Rupture energy of composite marble at different polyester content

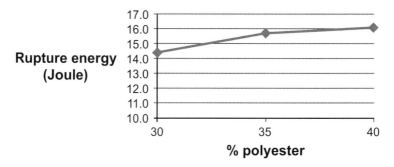

Figure 8.12. Rupture energy of composite slurry marble at different polyester content

Utilization of Marble and Granite Waste in Cement Manufacturing

Cement manufacturing is one of the best applications for marble and granite slurry waste in large quantities. However, the limestone as such does not possess hydraulic constituents in sufficient amounts; thus, development of strength cannot take place through reaction with water. It has to undergo a certain reaction and absorption. Calcination of limestone produces hydraulic lime. During the process of calcination, the oxides of silicon, aluminum, and iron present in situ react with lime and produce stable compounds of silicates/aluminates. These are the same that are present in Portland cement. These oxide compounds are hydraulic in nature, react with water, and become hard, as explained by Agarwal (2003). Studies show that it is technically possible to incorporate massive quantities of stone slurry as a raw material in the production of clinker without previous complex treatments. The Portuguese cement industry was a pioneer in this regard, as highlighted in

Almeida, Branco, and Santos (2007): the cement industry in Portugal obtains 3.5% of the total limestone raw material needed by the national cement industry from natural stone slurry. Nevertheless, the solution has not yet been generally adopted in spite of its technical viability. The viability of incorporating marble slurry as a replacement of limestone in the cement process in the town of Abu Road in the Sirohi District in Rajasthan is demonstrated in Sharma (2003). The cement plant requires 1.4–1.6 tons of limestone to produce one ton of clinker. The total generation of marble slurry works out to be more than 2% of the limestone required by the cement plants in Rajasthan.

A laboratory analysis comparing marble slurry and limestone is detailed in table 8.9. It is worth mentioning, however, that there are some constraints in the use of marble slurry, as indicated in Sharma (2003). These constraints are the high moisture content of the marble slurry (above 8%), which would pose a problem in feeding; the high content of magnesium (above 2.5%), which creates high linear expansion; the fineness of marble slurry, which is very high compared to the limestone used; the grain size of marble above 150 microns, which poses problems of compatibility during calcination, resulting in increased generation of fines in the product and requiring mineralizer (additional cost); and the unavailability of organized space for storage of slurry and the transportation cost from the generation point to the user plant. Yet

Table 8.9. Comparison between limestone and marble slurry for use in cement industry (Sharma, 2003)

Parameter (%)	Limestone	Marble slurry
SiO_2	12.6 ± 0.2	3.33 ± 0.2
Al_2O_3	3.5 ± 0.1	0.55 ± 0.1
Fe_2O_3	1.9 ± 0.1	0.54 ± 0.1
CaO	42.7 ± 0.2	49.11 ± 0.2
MgO	1.4 ± 0.05	5.0 ± 0.05
Na_2O	0.2 ± 0.05	0.15 ± 0.05
K_2O	0.45 ± 0.05	0.03 ± 0.05
SO_3	0.2 ± 0.05	0.43 ± 0.05
Moisture	0.5	12–18
Grain size, micron	80–90	180

these constraints can be solved by establishing a stockyard for marble slurry to control the disposal from the central point, a dewatering arrangement at the generation point before shifting to the central yard, and free transportation from waste generating units to the slurry plants. Technology for cleaner processing is available that could even be economically beneficial. It involves the installation of filter presses that separate water from the slurry, thus reducing the cost of transportation of this material. This water could then be recycled—not a bad economic concept in a water-scarce region.

Sampling and testing of cement

Five percent marble and granite slurry, powder, and mud were added to the clinker in the cement manufacturing process. Samples were produced in prisms measuring 40mm x 160mm, cut into 40mm cubes. The physical, chemical, and mechanical properties of the raw material and cement produced were determined and compared to ASTM C 150: Standard Specification for Portland Cement for Type I and EN 197 and ES 4756 for cement CEM I 32.5 N.

Chemical analysis

Table 8.10 shows the chemical analysis of the raw material used in the production and manufacturing of cement: limestone and clay, along with the proposed waste made up of marble powder, granite powder, and marble and granite in slurry form (mud). Marble powder has calcium oxide as the major component (> 50%), with loss of ignition (LOI) around 40%, and small amounts of SiO_2, MgO, and Fe_2O_3 (< 1%). By contrast, granite has SiO_2 as its major component (> 60%), with much lower levels of LOI (< 2%), with Al_2O_3 values between 11% and 15%, CaO values of 1% to 3.5%, and traces of Na_2O and K_2O (0 to 4.5%). The chemical analysis of granite slurry resulting from cutting operations using gang saws showed a higher value of Fe_2O_3 (14%), which reflects the use of iron grit in the cutting procedure as abrasive material.

The raw material analysis presented in table 8.10 coincides with the chemical analysis of marble slurry in waste indicated in table 8.9. Table 8.10 shows much lower levels (< 0.2% in marble sample), except for MgO (5.0 ± 0.05 in table 8.9). More importantly, the marble powder data matches the chemical composition of the raw materials used in manufacturing Portland cement (limestone and clay), which makes it a good alternative. Granite powder can replace limestone, in spite of the different chemical composition. Marble powder was successfully added to the clinker at a level of 5% without causing a major chemical change. However, granite powder and mud increased the insoluble residue (IR) to above 0.75%, 2.64%, and 3.44% respectively, which compromises the quality of cement if added to the clinker, according to ASTM C150 Type I specifications. On the other hand, according to EN 197 and ES 4756 for CEM I 32.5, IR can reach 5%. As for MgO and SO_3, all samples comply with ASTM C150 Type I, EN 197, and ES 4756 for CEM I 32.5, where the limits are 6 and 5 in clinker for MgO, and 3 and 3.5 for SO_3, respectively. The granite and mud samples failed to comply with the standards in the specified value of equivalent alkalis (less than 0.6). All samples show chlorine values less than 0.1 as specified in EN 197 and ES 4756 for CEM I 32.5. Loss of ignition (LOI) for all samples complies with EN 197 and ES 4756 for CEM I 32.5 (LOI < 5). However, the 5% limestone samples and the 5% marble samples fail to comply with ASTM C150 Type I (LOI < 3). It is worth mentioning, however, that all wastes, including granite powder, and slurry, can be used as a raw material for cement without causing any changes in the chemical properties of the cement.

Physical and mechanical analysis of cement
The physical analysis of cement is displayed in table 8.11. The results show that none of the physical properties were affected by the use of stone waste. The initial set time (> 75 minutes) and the expansion (< 10mm) are fulfilled according to EN 197 and ES 4756 for CEM I 32.5. All samples show acceptable values of compressive strength compared to those specified by the specifications mentioned earlier (10 N/mm^2 for 2-day test and 32.5 N/mm^2 for 28-day test). In addition, the 90-day tests for compressive strength demonstrate a plateau and prove that there was no deterioration in strength over time (figures 8.13, 8.14, 8.15, 8.16, 8.17).

Utilization of Marble Waste in Production of Stone Paper
The recycling process makes use of the extracted calcium carbonate, the major chemical component of marble waste, by mixing it with polyethylene and other additives. The mixture is extruded, forming sheets of stone paper. The extrusion process is done by melting a solid polymer and forcing it through a die by means of a screw. The die shapes the mixture into a final product such as a pipe, a sheet, or film.

Table 8.10. Chemical analysis of raw material and manufactured cement

Description	Limestone	Marble powder	Granite powder	Mud
SiO_2	1.35	0.71	62.44	66.98
Al_2O_3	0.1	0.1	11.46	15.08
Fe_2O_3	0.01	0.05	13.99	4.91
CaO	55.04	55.43	3.38	1.86
MgO	0.1	0.19	0.09	0.73
SO_3	0	0.04	0.02	0.01
Na_2O	0	0.01	2.93	3.61
K_2O	0	0	3.45	4.4
Equ. alkalis				
Cl	0	0.02	0.06	0.07
Mn_2O_3			0.14	0.16
TiO_2			0.37	0.36
P_2O_5			0	0
IR				
LOI	43.24	43.37	1.59	1.77
FL				
LSF	1409.66	2592	1.71	0.89
SM	12.27	4.73	2.45	3.35
AM	10	2	0.82	3.07
C_3S				
C_2S				
C_3A	0.25	0.18	6.71	31.62
C_4AF	0.03	0.15	42.54	14.92

Clay (comparison)	Blank (ck. + gypsum)	5% limestone	5% marble	5% granite	5% mud
50.54	20.3	19.56	19.7	19.42	19.51
15.77	4.82	4.61	4.67	4.99	5.21
9.69	4.72	3.81	3.74	4.51	4.07
7.76	63.12	62.81	62.54	61.04	60.31
4.01	1.53	1.39	1.44	1.44	1.47
0.09	2.81	2.8	2.86	2.87	2.75
1.23	0.19	0.14	0.2	0.33	0.42
0.8	0.05	0.45	0.44	0.59	0.68
	0.22	0.43	0.49	0.71	0.87
0.68	0.04	0.04	0.04	0.04	0.04
	0.33	0.21	0.22	0.23	0.21
	0.48	0.45	0.46	0.47	0.47
	0.19	0.15	0.14	0.16	0.15
	0.2	0.32	0.22	2.64	3.44
9.4	1.21	3.26	3.34	1.28	1.28
	1.23	1.06	1.12	1.23	1.23
4.66	93.21	97.09	95.91	93.43	92.07
1.99	2.13	2.32	2.34	2.04	2.1
1.63	1.02	1.21	1.25	1.1	1.28
	55.47	62.63	59.9	52.77	48.57
	16.37	8.83	11.31	15.87	19.3
25.39	4.79	5.76	6.05	5.58	6.91
29.46	14.35	11.59	11.38	13.72	12.37

Table 8.11. Physical analysis of cement

Description	Blank (ck. + gyp.)	5% limestone	5% marble	5% granite	5% mud
Blaine cm²/g	3236	3203	3220	3207	3223
Initial set (min)	115	145	160	150	135
Final set (hr)	2:50	3:20	3:35	3:25	3:10
Exp.	0	0	0	1	1
Sieve > 212	0.4	0.8	0.8	0.8	0.8
Sieve > 90	4.8	6.4	7.2	7.6	7.2
Sieve > 45	12	12.8	12.8	13.2	14
Sieve > 38	2.8	3.6	2.8	2.4	2.4
% Pass	80	76.4	76.4	76	75.6
% water cons.	24	23.75	23.75	24	24

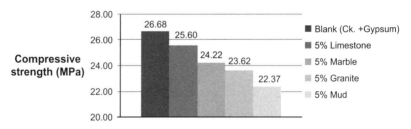

Figure 8.13. Compressive strength of cement samples after 2 days

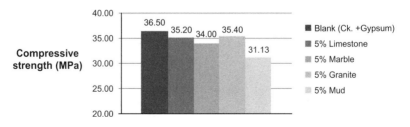

Figure 8.14. Compressive strength of cement samples after 7 days

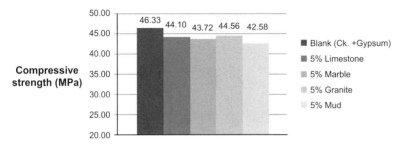

Figure 8.15. Compressive strength of cement samples after 28 days

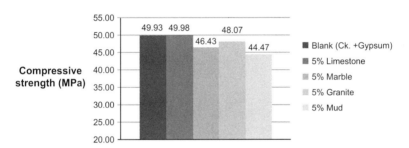

Figure 8.16. Compressive strength of cement samples after 60 days

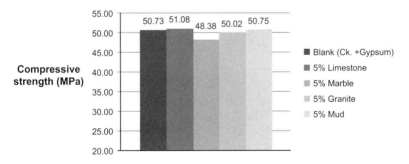

Figure 8.17. Compressive strength of cement samples after 90 days

The extrusion process also includes mixing the plastic with other additives such as coloring agents, reinforcing fibers, or adhesives. The extruder distributes all the mixture constituents equally and provides constant temperature and pressure throughout the mixture.

The stone paper sheets have a variable thickness, which is adjusted by means of a rolling system. The rolling system regulates the sheet drawing speed; the faster the drawing speed, the thinner the sheet, and vice versa. The sheet thickness varies from 180μm to 350μm depending on the drawing speed. The varying thicknesses allow the stone paper to be used for a large number of applications. The sheets with thickness ranging from 80μm to 200μm can be used for calendars, maps, instruction booklets, magazines, envelopes, tags, posters, handbags, plastic flooring, wallpaper, packaging paper, boards, advertising posters, or lightboxes. Sheet thicknesses ranging from 200μm to 500μm can be used for tags, greeting cards, name cards, brand cards, book covers, decorative boxes, and package boxes.

Utilization of Marble Waste in Production of Water-based Paints

Limestone is used in the production of paints, both as filler and as pigment. The preferred limestone for this purpose must contain more than 90% $CaCO_3$ and less than 1% impurities and a particle size of less than 2 microns. The average $CaCO_3$ content of the marble sludge is comparable to that required for the production of water paints. White sludge is preferred, and is carefully selected, since dark sludge can affect the whiteness of the paint. A great deal of marble sludge could be put to good use in this process (Mashaly, Shalaby, and El-Hefnawi, 2012).

Additives in the Plastic Industry

Polypropylene

To produce polypropylene, varying amounts of fillers are mixed together, depending on the requirements of the finished products. A particle size of less than 2 microns and a filler percentage of 30–40% is ideal for packaging films (Mashaly, Shalaby, and El-Hefnawi, 2012).

PVC

In PVC manufacture, $CaCO_3$ is largely used as filler, up to an amount of 60%. The particle size depends on the finished product, but the upper limit is 40 microns. If marble sludge is to be used as a filler, it must undergo grinding and drying before use (Mashaly, Shalaby, and El-Hefnawi, 2012) .

Utilization of Marble Waste in Production of Fertilizers

Calcium carbonate can be used in the production of fertilizer by a direct reaction with a solution of nitric acid to produce calcium nitrate $(Ca(NO_3)_2)$ fertilizer. The particle size of marble sludge seems to be fine enough (under 300 microns) for the intended use, and the moisture content of filter-pressed materials is acceptable since the reaction takes place in the aqueous state (Mashaly, Shalaby, and El-Hefnawi, 2012).

Desulfurization of Exhaust Gases

Calcium carbonate is also used to reduce sulfur oxide emissions produced as a result of fuel combusion in thermoelectric power plants. These emissions are harmful both to health and to the environment. They can be combined with $CaCO_3$ to produce gypsum. Marble sludge can be used without any restrictions as to its composition or humidity (Mashaly, Shalaby, and El-Hefnawi, 2012).

Production of Animal Food

Calcium carbonate can contribute up to 7% to 10% of the components of animal food. It must contain no heavy metals or toxic material, and have a suitable particle size and low moisture content. In addition, the calcium content must be at least 38%. White marble sludge can be used, provided it satisfies these criteria (Mashaly, Shalaby, and El-Hefnawi, 2012).

Production of Bituminous Mixes

Bituminous mixes are composed of approximately 6% bitumen and 94% inert materials. They are used to pave roads. Only dry marble dust with a large particle size can be used for this purpose. The extra energy costs for drying must be also be taken into account (Mashaly, Shalaby, and El-Hefnawi, 2012).

Production of Resin Products

Marble resin decorative plates are manufactured by mixing calcium carbonate composed of 25% white dust with grain size less than 200 microns, and 75% with grain size between 0.6 and 0.35mm, with polyester resin. The grain size of marble sludge is suitable for this process, but the water content should be adjusted to no more than 6–7% (Mashaly, Shalaby, and El-Hefnawi, 2012).

Recovery of Lead from Flat Batteries

Marble sludge, of any composition, size, or moisture content, can be used in the recovery of lead from car batteries by replacing soda ash (Mashaly, Shalaby, and El-Hefnawi, 2012).

Additives in Iron and Steel Manufacture

Marble sludge can be used in iron and steel manufacture by adding flux (limestone) to facilitate smelting and the formation of scraps or slag. The chemical composition of marble sludge is compatible with that of the flux. However, it requires treatment to adjust the particle size and lower the humidity content (Mashaly, Shalaby, and El-Hefnawi, 2012).

Production of Ceramics

The use of marble sludge in the production of structural ceramics is an option for reuse that is reported in the CORDIS "ReWaStone project" (1998). The process proved to be both technically and economically feasible, consuming a considerable amount of marble waste (Mashaly, Shalaby, and El-Hefnawi, 2012).

Production of Glass

Granite waste can be used in the production of glass bricks or other glass products because of its silica content. Marble powder can also be added in small quantities, at a ratio of 6:1. The mixture of granite and marble slurry is heated in a furnace until melting, with additives such as sodium hydroxide to reduce melting temperature, borax to add strength, and pigments, if required. The melted mixture is put into a press or blow mold to produce the glass product.

Summary

Marble and granite slurry cement bricks share similar mechanical properties, such as compressive strength, as well as physical properties, such as density and absorption. Granite slurry can be used in the production of brick samples, with an optimum amount of 10%. However, absorption is the major drawback of using slurry in cement bricks that are to meet the ASTM C55 standard, whose water-absorption requirement is fulfilled only at 0, 10%, and 20% slurry samples for grade S. The accelerated hydration, caused by heating, compensates for the detrimental effect of volumetric changes associated with temperature variation. Most cement brick samples, including the control, are of normal weight according to both the Egyptian specifications and ASTM C55. All cement brick samples tested in this study comply with the Egyptian code requirement for structural bricks, but not with ASTM C55. Instead, 10% and 20% marble and granite slurry yield Grade S. Most cement brick samples that contain marble and granite waste had sufficient abrasion resistance according to ASTM C902.

Composite marble could be manufactured from marble and granite waste. A mix design of 84% of marble waste and 16% polyester showed good physical and mechanical properties when compared with monolithic marble. Composite marble slurry samples (only powder and resin) give unsatisfactory results except for the rupture energy. It is not recommended for use in most marble applications.

The chemical analysis of marble powder, granite powder, and marble slurry (mud) proved that stone waste can be an alternative raw material in the manufacturing of cement, as compared with the conventional raw material used (limestone and clay), without causing any changes in the chemical properties of cement. Moreover, adding 5% marble and granite slurry, powder, and mud to the clinker in the cement manufacturing process did not reduce the quality of cement with respect to ASTM C150, EN 197, and ES 4756.

Marble powder can be used in dozens of applications, utilizing its major component of calcium carbonate. It can be used to produce stone paper, incorporating 60% polyethylene and 40% slurry. The paper thickness ranges from 180μm to 300μm. The paper can be used for plastic bags, calendars, envelopes, shredded curtains, and so on. Marble powder can be used as a filler in various products such as water paints, polypropylene, and PVC. It can be used as a source of calcium in manufacturing fertilizers and animal food. It can be used to desulfurize fumes from thermoelectric plants, in bituminous mixes after drying, in marble resin products, in recovery of lead from flat batteries, and as a catalyst in the iron and steel industry. Marble sludge can also be used in massive amounts in the manufacturing of ceramics.

Granite powder can be used to produce glass, owing to its high content of silica, the main constituent in glass production. The glass produced is of high quality compared to glass produced from raw sand.

Questions

1. Explain marble and granite production processes, the waste produced from each stage, its quantities, and its environmental impact.
2. Illustrate the physical and chemical characteristics of marble and granite waste produced as result of production processes.
3. Explain the effect of using marble and granite waste in concrete bricks, and investigate the physical and mechanical properties of the bricks produced.
4. Investigate the technical and economic feasibility of utilizing marble waste in producing cultured marble.
5. Study the technical feasibility of incorporating marble and granite slurry in the manufacturing of cement.
6. List possible advantages of utilizing marble and granite waste in different industries, along with the constraints and limitations.
7. Investigate the constraints of using marble and granite slurry in different industries, along with possible solutions.

References

Agarwal, S. 2003. "Waste: A Gateway to the Future Economy of Kotah Stone Industry." Centre for Development of Stones. www.cdos-india.com/Papers%20technical.htm

Almeida, N., F. Branco, and J.R. Santos. 2007. "Recycling of Stone Slurry in Industrial Activities: Application to Concrete Mixtures." *Building and Environment* 42: 810–19. www.sciencedirect.com

Binici, H., H. Kaplan, and S. Yilmaz. 2007. "Influence of Marble and Limestone Dusts as Additives on Some Mechanical Properties of Concrete." *Scientific Research and Essay* 2, no. 9: 372–79. www.academicjournals.org/SRE

Borsellino, C., L. Calabrese, and G. Di Bella. 2008. "Effects of Powder Concentration and Type of Resin on the Performance of Marble Composite Structures." *Construction and Building Materials* 23, no. 5: 1915–21. www.sciencedirect.com

Celik, M., and E. Sabah. 2008. "Geological and Technical Characterisation of Iscehisar (Afyon, Turkey) Marble Deposits and the Impact of Marble Waste on Environment Pollution." *Journal of Environmental Management* 87: 106–16. www.sciencedirect.com

Ciccu, R., R. Cosentino, C.C. Montani, A. El Kotb, and H. Hamd. 2005. "Strategic Study on the Egyptian Marble and Granite Sector." Industrial Modernisation Centre. www.imc-egypt.org/en/studies/Strategic%20study%20on%20the%20Egyptian%20Marble%20and%20Granite%20Sector.pdf

Corinaldesi, V., G. Moriconi, and T. Naik. 2010. "Characterization of Marble Powder for Its Use in Mortar Concrete." *Construction and Building Materials* 24: 113–17. www.sciencedirect.com

Delgado, J., A. Vázquez, R. Juncosa, and V. Barrientos. 2006. "Geochemical Assessment of the Contaminant Potential of Granite Fines Produced during Sawing and Related Processes Associated to the Dimension Stone Industry." *Journal of Geochemical Exploration* 88: 24–27. www.sciencedirect.com

El-Haggar, S. 2007. *Sustainable Industrial Design and Waste Management: Cradle-to-Cradle for Sustainable Development.* San Diego: Elsevier Academic Press.

Hamza, R., S. El-Haggar, and S. Khedr. 2011a. "Marble and Granite Waste: Characterization and Utilization in Concrete Bricks." *International Journal of Bioscience, Biochemistry and Bioinformatics* 1: 286–91. www.ijbbb.org/papers/54-B108.pdf

Hamza, R., S. El-Haggar, and S. Khedr. 2011b. "Utilization of Marble and Granite Waste in Concrete Brick." 2011 International Conference on Environment and BioScience. *International Journal of Bioscience, Biochemistry and Bioinformatics* 21. www.ipcbee.com/vol21/22-ICEBS2011P00013.pdf

Kandil, A., and T. Selim. 2007. "Characteristics of the Marble Industry in Egypt." www.aucegypt.edu/academics/dept/econ/Documents/marble.pdf

Kumar, M., K. Suryanarayana, and T. Venkatesh. 2003. "Value Added Products from Marble Slurry Waste." Centre for Development of Stones. www.cdos-india.com/Papers%20technical.htm

Mashaly, A., B. Shalaby, and M. El-Hefnawi. 2012. "Characterization of the Marble Sludge of the Shaq El Thoaban Industrial Zone, Egypt and Its Compatibility for Various Recycling Applications." *Australian Journal of Basic and Applied Sciences* 6, no. 3: 153–61. www.ajbasweb.com/ajbas/2012/March/153-61.pdf

Menezes, R., H.S. Ferreira, G. Neves, H. Libra, and H.C. Ferreira. 2005. "Use of Granite Sawing Wastes in the Production of Ceramic Bricks and Tiles." *Journal of the European Ceramic Society* 25: 1149–58. www.sciencedirect.com

Minerals Zone. 2005. "Marble." www.mineralszone.com/stones/marble.html

Misra, A., and R. Mathur. 2003. "Marble Slurry Dust (MSD) in Roads and Concrete Work." Centre for Development of Stones. www.cdos-india.com/Papers%20technical.htm

Monteiro, S., L. Pecanha, and C. Vieira. 2004. "Reformulation of Roofing Tiles Body with Addition of Granite Waste from Sawing Operations." *Journal of the European Ceramic Society* 24: 2349–56. www.sciencedirect.com

Pareek, S. 2007. "Gainful Utilization of Marble Waste: An Effort towards Protection of Ecology and Environment." GOGO Stone. http://optimum1.gogostone.cn/newsinfo.asp?nic=430

Rizzo, G., F. D'Agostino, and L. Ercoli. 2008. "Problems of Soil and Groundwater Pollution in the Disposal of 'Marble' Slurries in NW Sicily." *Environmental Geology* 55: 929–35. www.sciencedirect.com

Saboya, F., G. Xavier, and J. Alexandre. 2007. "The Use of Powder Marble By-product to Enhance the Properties of Brick Ceramic." *Construction and Building Materials* 21: 1950–60. www.sciencedirect.com

Sharma, N. 2003. "Presentation on Use of Marble Slurry as a Part Substitute of Limestone in a Cement Plant." Centre for Development of Stones. www.cdos-india.com/Papers%20technical.htm

Torres, P., H. Fernandes, S. Olhero, and J. Ferreira. 2009. "Incorporation of Wastes from Granite Rock Cutting and Polishing Industries to Produce Roof Tiles." *Journal of the European Ceramic Society* 29: 23–30. www.sciencedirect.com

Vijayalakshmi, V., S. Singh, and D. Bhatnagar. 2003. "Developmental Efforts in R&D for Gainful Utilization of Marble Slurry in India." Centre for Development of Stones. www.cdos-india.com/Papers%20technical.htm

CHAPTER 9

Beyond Sustainability for the Production of Fuel, Food, and Feed

Dalia Nakhla and Salah M. El-Haggar

Introduction

Sustainable development was first defined in 1987 by the United Nations as "development that meets the needs of the present without compromising the ability of future generations to meet their own needs" (UN, 1987). This definition contains within it two concepts: that of 'needs,' and that of the limitations on the environment's ability to meet present and future needs. The United Nations 1992 Conference on Environment and Development, usually referred to as the Rio (or Earth) Summit, resulted in the Rio Declaration on Environment and Development, or Agenda 21. Sustainability was defined as the combination of technological, scientific, environmental, and social systems in such a way that the resulting systems can be retained in a state of temporal and spatial equilibrium (Roosa, 2010).

"Beyond sustainability," on the other hand, goes beyond the preservation of existing resources. It is a concept that encourages creativity and innovation with the aim of generating new resources from wastes, effluents, and emissions. "Beyond sustainability" views pollution as an opportunity rather than a burden on the environment. Wastes, effluents, and emissions could be recovered and transformed into raw material and resources that could take the place of our natural resources. An example of this is the promotion of the growth of algae by absorbing carbon dioxide emissions from stacks of industrial establishments and power plants. Algae can also help to address the issues of water pollution and contamination, as they can flourish on waste water, removing the excess nutrients and absorbing the heavy metals. The cultivated algae that have been used for removing pollution have become a valuable innovative resource that could be used as food, fodder, or fuel.

Fuel, food, and feed are our three main sources of energy. While fuel gives machines and devices the needed energy to operate, food gives humans the energy for their bodies to operate. Feed is an indirect source of food, as feed is consumed by animals, birds, and fish, which are in turn consumed by humans. The increasing price of oil and scarcity of its production are pushing the world to find alternative new fuel sources. Every day there is extraordinary growth in industrialization and population; hence, the desire for energy continues to increase.

The traditional sources of energy are coal, oil, and natural gas. Yet dependence on fossil fuels has severe impacts on the environment, as fossil fuel combustion pollutes the atmosphere with its associated greenhouse gas (GHG) emissions. In addition to GHG emissions, petroleum combustion is the main source of other air pollutants, including nitrogen oxides (NO_x), sulfur oxides (SO_x), carbon oxides (CO_x), particulate matter, and volatile organic compounds.

Energy recovered from biomass is a crucial sustainable source of fuel because it mitigates greenhouse gas emissions and acts as an alternate for non-renewable fossil fuels. Therefore, large-scale establishment of biomass energy in a form of biofuel could contribute to sustainable development on several fronts: environmental, social, and economic.

Biofuel is obtained from renewable biomass sources such as waste cooking oil derived from palm, soybean, canola, rice bran, sunflower, coconut, corn, fish, chicken, and algae. It is non-hazardous and biodegradable, and reduces reliance on petroleum-based fuel.

One of the major contributors to global warming has been the increase in CO_2 concentrations in the atmosphere resulting from the burning of fossil fuels. Biomass, by contrast, fixes CO_2 in the atmosphere through photosynthesis, and is thus considered an environmentally friendly energy source. Sustainable use of biomass can be CO_2-neutral, because the CO_2 released by the burning of biomass is counterbalanced by the CO_2 fixed by photosynthesis.

Algae (both macro- and microalgae) usually have superior photosynthetic efficiency over other types of biomass. They can generate up to 250 times the amount of oil per acre as soybeans (Chisti, 2007). In fact, generating biofuel from algae may be the only way to produce enough automotive fuel to substitute current fossil usage. Algae produce 7 to 31 times more oil than palm oil (Chisti, 2007). It is very easy to extract oil from algae. The idea of using microalgae as a source of alternative fuel is not new, but it is now being taken more seriously because of the escalating price of petroleum and, more importantly, the rising concern about global warming associated with burning fossil fuels.

Algae, Humanity's Best Friend
Algae can be considered "humanity's best friend" (Edwards, 2009), as they can provide sustainable and affordable food and fuel, as well as ecological solutions.

Land plants evolved from algae five billion years ago. Therefore, any food, fibers, or materials that can be made from land-based crops can be made from algae. Moreover, fuels, plastics, or other materials made from fossil fuels can be made from algae, since fossil fuels are basically fossilized algae or the organisms that consume algae.

Algae can withstand and even take advantage of the main environmental problems that the earth is currently facing, including greenhouse gases, global warming, and land, water, and soil pollution. Algae production can recycle or consume CO_2 from the atmosphere. Food crops will suffer from global warming, but algae prosper in heat. Moreover, unlike food crops, algae can produce food and energy on non-cropland.

In the absence of fresh water, algae can flourish in waste, brine, or ocean water. "Algal cultivation can produce valuable biomass using no or minimal fossil resources that compete with land-based food crops, and do not require fertile soils, freshwater, fossil fuels, fertilizers, and fossil agricultural chemicals" (Edwards, 2009).

Algae not only survive in a polluted environment, but could also help reduce pollution if they are located near carbon sources such as power plants, cement plants, or breweries. In the process of cleaning water and air, the algae transform CO_2 and waste nutrients to valuable sugars, proteins, lipids, carbohydrates, and other organic compounds. Algae can produce carbon-neutral food and fuel with a positive ecological footprint. Algal fuels burn cleanly because they are basically vegetable oil (Edwards, 2009).

History of Algae Use
Oxygen
Studies show that the earth in its earliest form did not support life or food. Its temperature was extremely high, it had no oxygen, and it was covered by carbon dioxide gas. The essential role of algae emerged billions of years ago, converting the world to a more livable place. That radical change, which prepared the earth's environment for living things, came about as a result of the algae's

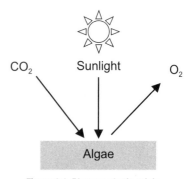

CO_2 Sunlight O_2

Algae

Figure 9.1. Photosynthetic origin of atmospheric oxygen

utilization of the energy from the sun in a photosynthesis reaction, breaking the carbon dioxide bond to produce free oxygen. Fossil analysis shows that the first living creatures on earth were cyanobacteria, the nano-particle-sized blue-green algae responsible for this early transformation of the earth's atmosphere. The work of the microalgae proceeded very slowly, molecule by molecule, for almost 2.5 billion years, until the inhospitable carbon dioxide environment was completely converted to an oxygen environment capable of satisfying life's needs (fig. 9.1). Only after this major renovation of the earth's environment was the emergence of plants, reptiles, fish, and other living creatures made possible.

Despite their small size, algae play a major part in oxygen production, much more than all the forests of the world put together (Edwards, 2009).

Early food
Algae had to go through many hardships, evolving millions of times in response to the destruction of their micro-environments as a result of the harsh surrounding conditions of intolerable heat, wind, or freezing weather. They managed to survive, however, developing many resistance techniques that made their growth and multiplication possible. The species persistence also led to the evolution of millions of algal species, all different in composition.

Adding to the unrivalled benefit of oxygen, algae themselves were the source of food for many of the early living creatures. Since algae are the lowest in the food chain, early life depended heavily on them as a food source. As a result, their survival approaches evolved not only to resist the harsh environments, but also to overcome the heavy consumption by other species. They were forced to grow faster than this consumption rate and faster than all other species. The remarkable ability of algae to produce millions of offspring a day from one cell allowed them to almost double their mass each day, adding to their survival competitiveness (Edwards, 2009).

Types of Algae

Algae are divided into two groups: phytoplankton (microalgae) and macrophytes (macroalgae). Microalgae are unicellular green plants rich in chlorophyll that lack lignin or cellulose and contain proteins, carbohydrates, and lipids in strain-specific proportions. They are plentiful in oceans, suitable for cultivation in rivers, and serve as the primary source of carbohydrates and proteins for aquatic organisms (Velasquez-Orta, Curtis, and Logan, 2009). They are prokaryotic or eukaryotic photosynthetic microorganisms that can grow rapidly and live in harsh conditions due to their unicellular or simple multicellular structure. The three most important classes of microalgae in terms of abundance are the diatoms (Bacillariophyceae), the green algae (Chlorophyceae), and the golden algae (Chrysophyceae) (Demirbas and Demirbas, 2010). The cyanobacteria or blue-green algae (Cyanophyceae) are also referred to as microalgae (Demirbas and Demirbas, 2010).

Macroalgae, or 'seaweeds,' are multicellular plants growing in salt or fresh water (Demirbas and Demirbas, 2010). They are more resistant to predators and environmental conditions than microalgae and are abundant in coastal zones. They lack lignin, and largely consist of polysaccharides (alginate, laminaran, and mannitol) and unsaturated fatty acids which are easily hydrolyzed, and low concentrations of cellulose (Velasquez-Orta, Curtis, and Logan, 2009). They are often fast-growing and can reach sizes of up to 60 meters in length. They are classified into three broad groups based on their pigmentation: brown seaweed (Phaeophyceae), red seaweed (Rhodophyceae), and green seaweed (Chlorophyceae) (Demirbas and Demirbas, 2010).

Both macrophytes and phytoplankton are suitable for cultivation using river water, seawater, and some wastewaters (Velasquez-Orta, Curtis, and Logan, 2009). Macro- and microalgae are currently used for food, in animal feed, in feed for aquaculture, and as biofertilizer. Seaweeds are used in the production of food, feed, chemicals, cosmetics, and pharmaceutical products (Demirbas and Demirbas, 2010).

Most microalgae need light and carbon dioxide as energy and carbon sources. This culture mode is usually called 'photoautotrophic.' Some algae species, however, are capable of growing in darkness and of using organic carbons (such as glucose or acetate) as energy and carbon sources. This culture mode is termed 'heterotrophic' (Demirbas and Demirbas, 2010).

Algae for Fuel

Biomass can be used in the production of renewable sources of energy. Though there are many sources of biomass that can contribute to the production of oil, such as waste cooking oil, oil crops, and animal fat, they cannot be regarded as practical alternative sources of fuel in the future. This is due to the biological fact that the amount of oil they can produce per hectare is low; large land spaces would be required in order to satisfy the extraordinary amount of fuel consumed just by the transportation sector. These large land spaces may not be available. For example, if the oil palm were used as an alternative renewable source for fuel, it would consume 24% of oil-palm cultivation, which would affect the other uses of oil-palm crops (Chisti, 2007), as illustrated in table 9.1.

As table 9.1 shows, algal biomass is the most practical way to the sustainable production of fuel due to its high oil content, which requires cultivation of minimal areas of lands (Chisti, 2007).

Biofuels

Biofuels are solid, liquid, or gaseous fuels that are produced from biorenewable or combustible renewable feedstocks. In the future, petroleum fuels could be replaced with liquid biofuels such as bioethanol and biodiesel. The biggest difference between biofuels and petroleum feedstocks is oxygen content. Biofuels are non-polluting, locally available, accessible, sustainable, and reliable fuels obtained from renewable sources. Biomass-based energy sources for heat, electricity, and transportation fuels are potentially carbon-dioxide neutral and recycle the same carbon atoms. Table 9.2 shows the major benefits of biofuels (Demirbas and Demirbas, 2010).

Biofuels can be classified based on their production technologies: first-generation biofuels, second-generation biofuels, third-generation biofuels, and fourth-generation biofuels. They can be described as follows.

Table 9.1. Comparison of oil yield from various sources of biomass (Chisti, 2007)

Crop type	Oil yield (L/ha)	Land area required (Mha)[a]	Percent of existing US crop area[a]	Percent of existing US crop area[b]
Corn	172	1,540	846	1,692
Soybean	446	549	326	652
Canola	1,190	223	122	244
Jatropha	1,892	140	77	154
Coconut	2,689	99	54	108
Oil palm	5,950	45	24	48
Microalgae[c]	136,900	2	1.1	2.2
Microalgae[d]	58,700	4.5	2.5	5

[a] To meet 50% of all transport fuel needs in the United States
[b] To meet 100% of all transport fuel needs in the United States
[c] 70% oil (by weight) in biomass
[d] 30% oil (by weight) in biomass

1. First-generation biofuels are made from sugar, starch, vegetable oils, or animal fats using conventional technology. The basic feedstocks are usually seeds or grains. Wheat, for example, yields starch that is fermented into bioethanol; sunflower seeds are pressed to yield vegetable oil that can be used in biodiesel.

2. Second-generation biofuels are made from nonfood crops, wheat straw, corn, wood, and energy crops using advanced technology.

3. Third-generation biofuels are made from algae that give high yield for low input (producing 30 times more energy per acre than land) to produce biofuels using more advanced technology.

4. Fourth-generation biofuels consist of biogasoline made from vegetable oil and biodiesel, using the most advanced technology.

Table 9.3 shows the classification of renewable biofuels based on their production technologies.

Table 9.2. Major benefits of biofuels (Demirbas and Demirbas, 2010)

Economic impacts	Sustainability
	Fuel diversity
	Increased number of rural manufacturing jobs
	Increased income taxes
	Increased investments in plant and equipment
	Agricultural development
	International competitiveness
	Reduced dependency on imported petroleum
Environmental impacts	Greenhouse gas reductions
	Reduced air pollution
	Biodegradability
	Higher combustion efficiency
	Improved land and water use
	Carbon sequestration
Energy security	Domestic targets
	Supply reliability
	Reduced use of fossil fuels
	Ready availability
	Domestic distribution
	Renewability

Oil production process

In the presence of sunlight, carbon dioxide, and organic nutrients present in water, algae can photosynthesize organic matter and double their mass. Depending on the species, about half their mass is made of lipids or natural oils. These can be extracted and used as "straight algal crude" or "refined to higher grade hydrocarbon products" such as biodiesel and biojet fuel. The types of algae that produce more carbohydrates than oil can be fermented to make bioethanol and biobutanol (Marsh, 2009).

Microalgae have been tested as raw material for the production of bio-oil, methane, methanol, and hydrogen. Macroalgae, on the other hand, have primarily been used to produce methane. Fuel produced from these technologies must be stored, transported, and further processed to produce electricity (Velasquez-Orta, Curtis, and Logan, 2009).

Table 9.3. Classification of renewable biofuels according to their production technologies
(Demirbas and Demirbas, 2010)

Generation	Feedstock	Example
First	Sugar, starch, vegetable oils, animal fats	Bioalcohols, vegetable oil, biodiesel, biosyngas, biogas
Second	Non-food crops, wheat straw, corn, wood, solid waste, energy crops	Bioalcohols, bio-oil, bio-DMF, biohydrogen, bio-Fischer-Tropsch diesel, wood diesel
Third	Algae	Vegetable oil, biodiesel
Fourth	Vegetable oil, biodiesel	Biogasoline

Algae biofuels contain no sulfur, are non-toxic, and are biodegradable. Some strains produce fuel with the same energy density as conventional fossil fuels. They are carbon neutral, as the emissions resulting from burning the fuel are balanced by the absorption of CO_2 by the growing organisms (Marsh, 2009). Table 9.4 lists the advantages and disadvantages of biofuel production using microalgae.

Table 9.5 presents a comparison of microalgae with other biodiesel feedstock in terms of seed oil content and oil yield per hectare of land. As shown, even microalgae with low oil content exceed all other oil plants in oil content and yield, as well as in biodiesel.

Table 9.4. Advantages and disadvantages of biofuel production using microalgae
(Demirbas and Demirbas, 2010)

Advantages	Disadvantages
High growth rate	Low biomass concentration
Less water demand than land crops	Higher capital costs
High-efficiency CO_2 mitigation	
More cost-effective farming	
Minimization of nitrous oxide release	

Table 9.5. Comparison of microalgae with other biodiesel feedstock
(Mata, Martins, and Caetano, 2010)

Plant source	Seed oil content	Oil yield	Land use	Biodiesel productivity
	(% oil by wt in biomass)	(L oil/ha/year)	(m^2/year/kg biodiesel)	(kg biodiesel/ha/year)
Corn/maize (*Zea mays* L.)	44	172	66	152
Hemp (*Cannabis sativa* L.)	33	363	31	321
Soybean (*Glycine max* L.)	18	636	18	562
Jatropha (*Jatropha curcas* L.)	28	741	15	656
Camelina (*Camelina sativa* L.)	42	915	12	809
Canola/rapeseed (*Brassica napus* L.)	41	974	12	862
Sunflower (*Helianthus annuus* L.)	40	1070	11	946
Castor (*Ricinus communis*)	48	1307	9	1156
Palm oil (*Elaeis guineensis*)	36	5366	2	4747
Microalgae (low oil content)	30	58,700	0.2	51,927
Microalgae (medium oil content)	50	97,800	0.1	86,515
Microalgae (high oil content)	70	136,900	0.1	121,104

Figure 9.2 shows a schematic representation of the algal biodiesel value chain stages. The chain begins with the selection of microalgae species depending on local specific conditions and the design and implementation of the cultivation system for microalgae growth. It continues through biomass harvesting, processing, and oil extraction to supply the biodiesel production unit.

Another possibility is to power systems with integrated algae, as shown in figure 9.3. The process begins with providing water, inorganic nutrients, carbon dioxide, and light to microalgal culture during the biomass-production stage. This is followed by the biomass-recovery stage, where the cells suspended in the broth are separated from the water and residual nutrients, which are then recycled to the biomass-production stage. The recovered biomass is used for extracting the algal oil that is further converted to biodiesel in a separate process. The spent biomass can be used as animal feed and for recovering other possible high-value products that might be present in the biomass. However, the main bulk of the biomass undergoes anaerobic digestion, which produces biogas to generate electricity. Effluents from the anaerobic digester are used as a nutrient-rich fertilizer and as irrigation water. Most of the power generated from the biogas is consumed within the biomass-production process, and any excess energy is sold to grid. Carbon dioxide emissions from the power-generation stage are fed into the biomass production (Chisti, 2007).

Microalgal biomass production
Currently there are three types of industrial reactors used for algal culture: photobioreactors, open ponds, and hybrid systems (Demirbas and Demirbas, 2010).

Photobioreactors
Photobioreactors are basically tanks or closed systems where algae are cultivated. They produce algae while performing other beneficial tasks, such as scrubbing power plant flue gases or removing nutrients from wastewater (Demirbas and Demirbas, 2010). Photobioreactors offer a closed culture environment, which is protected from direct fallout and so is relatively safe from invading microorganisms. This technology is relatively expensive compared to the open ponds because of the infrastructure costs (Demirbas and Demirbas, 2010). However, unlike open ponds they do not suffer from issues of land use cost, water availability,

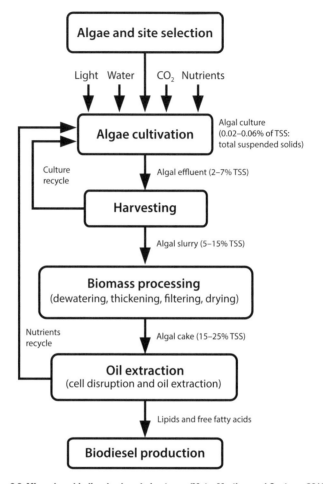

Figure 9.2. Microalgae biodiesel value chain stages (Mata, Martins, and Caetano, 2010)

and appropriate climatic conditions. Moreover, photobioreactors have higher efficiency and biomass concentration (2 to 5 g/L), shorter harvest time (2 to 4 weeks), and higher surface-to-volume ratio (25 to 125/m) than open ponds (Demirbas and Demirbas, 2010).

Tubular photobioreactors consist of an array of straight transparent tubes made of glass or plastic (Chisti, 2007). Tubes are placed in parallel to each other or flat above the ground to maximize the illumination surface-to-volume ratio of the reactor. Figure 9.4 shows a tubular photobioreactor with parallel-run horizontal tubes.

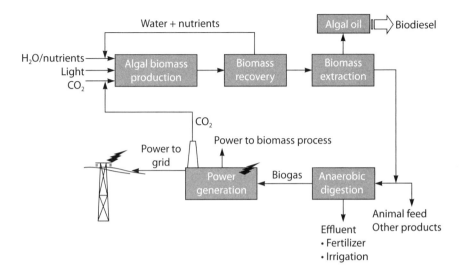

Figure 9.3. A conceptual process for producing microalgal oil for biodiesel (Chisti, 2008)

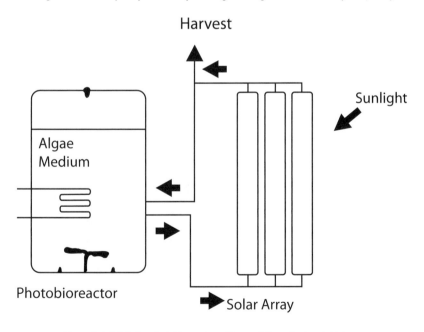

Figure 9.4. A tubular photobioreactor with parallel-run horizontal tubes
(Demirbas and Demirbas, 2010)

The diameter of the tubes is usually small (0.2m diameter or less) to allow light penetration to the center of the tube (Demirbas and Demirbas, 2010). The microalgal broth circulates from a reservoir to the reactor and back to the reservoir. A turbulent flow is maintained in the reactor to ensure distribution of nutrients, improve gas exchange, minimize cell sedimentation, and circulate biomass for equal illumination between the light and dark zones (Demirbas and Demirbas, 2010). Flow is produced using either a mechanical pump or a gentler airlift pump. Periodically, photobioreactors must be cleaned and sanitized. This is easily achieved using automated clean-in-place operations (Chisti, 2008).

As shown in figure 9.5, photosynthesis generates oxygen. Under high irradiance, oxygen can be generated in a typical tubular photobioreactor at a rate as high as 10g O_2 m^{-3} min^{-1}. If the dissolved oxygen levels exceed the air saturation values by too much, photosynthesis may be inhibited and the algal cells may suffer photooxidative damage. To prevent this, the maximum tolerable dissolved oxygen level should not generally exceed about 400% of air saturation value. To remove excess oxygen, the culture must periodically return to a degassing zone that is bubbled with air to strip out the accumulated oxygen (Chisti, 2008).

As the broth moves along a photobioreactor tube, the pH increases because of consumption of carbon dioxide. Carbon dioxide is fed into the degassing zone in response to a pH controller. Photobioreactors require cooling during daylight hours and temperature control at night. A heat exchange coil may be located in the degassing column, as shown in figure 9.5.

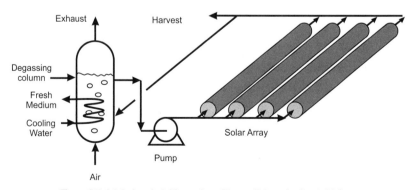

Figure 9.5. A tubular photobioreactor with parallel-run horizontal tubes (Demirbas and Demirbas, 2010)

Figure 9.6. (a) Tubular and (b) vertical photobioreactors (Demirbas and Demirbas, 2010)

Closed system bioreactors or photobioreactors (PBR) have the main advantage that the algal broth is safe from windborne contamination and evaporation. They also support up to five-fold higher productivity with respect to reactor volume, and thus have a smaller 'footprint' on a yield basis (Demirbas and Demirbas, 2010). Closed systems also save water, energy, and chemicals, and so have lower cost. They allow for single-species culture of microalgae for prolonged periods. The main problem with a closed system is the source of CO_2. This can be obtained from any industrial CO_2-emitting stack. Figure 9.6 shows the two main types of photobioreactors: tubular and vertical.

Open-pond systems

Open ponds are the oldest and simplest systems for mass cultivation of microalgae. The pond is designed in a raceway configuration (figure 9.7), in which a paddlewheel circulates and mixes the algal cells and nutrients. Flow is guided around bends by baffles placed in the flow channel. Raceway channels are typically made from concrete, or consist of an earth structure with a plastic liner. The system is often operated in a continuous mode, that is, the fresh feed is added in front of the paddlewheel, and the algal broth is harvested behind the paddlewheel after it has circulated through the loop. Nutrients can be provided through runoff water from nearby land areas or by channeling the water from sewage/water treatment plants (Demirbas and Demirbas, 2010). Although this system is the easiest and cheapest to construct, it may suffer from contamination, evaporation, temperature control, CO_2 utilization, and maintainability (Chisti, 2008).

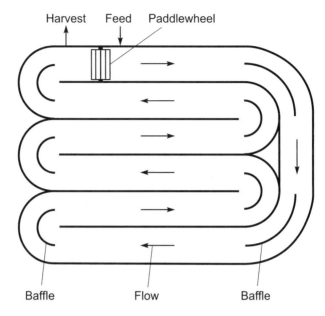

Figure 9.7. Aerial view of a raceway pond (Chisti, 2007)

Hybrid systems

Open ponds provide a suitable and profitable environment for algal growth, but they easily become contaminated with harmful species. In hybrid systems, a combination of open ponds and closed bioreactors is used. This is a cost-effective option for cultivating high-yielding strains for biofuels. Cleaning and flushing of ponds should be part of the aquaculture routine to minimize contamination. It is also essential to choose a site with the abundant sunlight that is necessary for photosynthesis. Light can be enhanced by solar collectors or solar concentrators (Demirbas and Demirbas, 2010). Table 9.6 compares open and closed large-scale culture systems for microalgae.

Harvesting and biomass concentration

Algal harvesting or biomass recovery can return 20–30% of the total biomass production cost. There is no one suitable method for harvesting; the most common include sedimentation, centrifugation, filtration, ultra-filtration combined with a flocculation step, or a combination of flocculation and flotation to increase particle size and enhance sedimentation (Mata, Martins, and Caetano, 2010).

Table 9.6. A comparison of open and closed large-scale culture systems for microalgae
(Mata, Martins, and Caetano, 2010)

Culture systems for microalgae	Closed systems (PBRs)	Open systems (ponds)
Contamination control	Easy	Difficult
Contamination risk	Reduced	High
Sterility	Achievable	None
Process control	Easy	Difficult
Species control	Easy	Difficult
Mixing	Uniform	Very poor
Operation regime	Batch or semi-continuous	Batch or semi-continuous
Space required	A matter of productivity	PBRs ≈ Ponds
Area/volume ratio	High (20–200 m/1)	Low (5–10 m/1)
Population (algal cell) density	High	Low
Investment	High	Low
Operation costs	High	Low
Capital/operating costs ponds	Ponds 3–10 times lower cost	PBRs > Ponds
Light utilization efficiency	High	Poor
Temperature control	More uniform temperature	Difficult
Productivity	3–5 times more productive	Low
Water losses	Depends upon cooling design	PBRs ≈ Ponds
Hydrodynamic stress on algae	Low–high	Very low
Evaporation of growth medium	Low	High
Gas transfer control	High	Low
CO_2 losses	Depends on pH, alkalinity, etc.	PBRs ≈ Ponds
O_2 inhibition	Greater problem in PBRs	PBRs > Ponds
Biomass concentration	3–5 times in PBRs	PBRs > Ponds
Scale-up	Difficult	Difficult

The four primary harvesting methods that have been found to be suitable for biofuel production are microstraining, belt filtering, flotation with float collection, and sedimentation. Microstrainers are preferred for their mechanical simplicity and availability in large unit sizes. Some studies recommend flocculation prior to microstraining. Filter press, on the other hand, can recover large quantities of biomass through application of pressure or vacuum but can be relatively slow. Moreover, filtration is better suited for large microalgae such as *Coelastrum proboscideum* and *S. platensis*, but cannot recover organisms with smaller dimensions such as *Scenedesmus*, *Dunaliella*, or *Chlorella*. Fragile cells and small-scale production can be handled through membrane microfiltration and ultra-filtration (Mata, Martins, and Caetano, 2010).

Sedimentation tanks or settling ponds are suited for recovering biomass from sewage-based processes. Centrifugation is more applicable to recover high-quality algae such as for food or aquaculture applications. Centrifuge devices can be easily cleaned or sterilized to effectively avoid bacterial contamination or fouling of the raw product (Mata, Martins, and Caetano, 2010).

The density of the biomass produced affects the choice of harvesting method, as gravity-sedimented sludge is more diluted than centrifugally recovered biomass, which influences the cost of thermal drying of the mass. In all cases, once separated, the algal biomass (5–15% dry weight) must be processed quickly, as it will spoil in only a few hours in a hot climate (Mata, Martins, and Caetano, 2010).

Processing and components extraction

Algae mass is processed to produce either low-cost commodities, such as fuels, feeds, and foods, or higher-value products, such as b-carotene and polysaccharides. The processing method depends on the desired product.

The biomass is usually dehydrated to increase its shelf life. This can be done through spray-drying, drum-drying, freeze-drying, or sun-drying, depending on the moisture content of the sludge (Mata, Martins, and Caetano, 2010). After drying, cell disruption is carried out to release the metabolites. These methods include mechanical processes (for example, cell homogenizers, bead mills, ultrasounds, autoclave, and spray-drying) and non-mechanical processes (such as freezing, organic solvents and osmotic shock, and acid, base, and enzyme reactions) (Mata, Martins, and Caetano, 2010).

For biodiesel production, lipids and fatty acids have to be extracted from the microalgal biomass. Lipids can be quickly and efficiently extracted directly from the lyophilized biomass using solvents to reduce degradation. Solvents include hexane, ethanol (96%), or a hexane–ethanol (96%) mixture, from which it is possible to obtain up to 98% quantitative extraction of purified fatty acids. Ethanol is recommended when the purpose is just to extract the lipids without cellular contaminants such as sugars, amino acids, salts, hydrophobic proteins, and pigments (Mata, Martins, and Caetano, 2010).

Extraction methods such as ultrasound- and microwave-assisted procedures have also been studied for oil extraction from vegetable sources. As compared with conventional methods, these new methods can greatly improve oil extraction with higher efficiency. Extraction times were reduced and yields increased by 50–500% with low or moderate costs and minimal added toxicity (Mata, Martins, and Caetano, 2010).

Biodiesel production

Biodiesel is a mixture of fatty acid alkyl esters obtained by transesterification (ester exchange reaction) of vegetable oils or animal fats. These lipid feedstocks are composed of 90–98% (by weight) of triglycerides and small amounts of mono- and diglycerides, free fatty acids (1–5%), and residual amounts of phospholipids, phosphatides, carotenes, tocopherols, sulfur compounds, and traces of water (Mata, Martins, and Caetano, 2010). Transesterification is a multiple-step reaction, including three reversible steps in series, where triglycerides are converted to diglycerides, then diglycerides are converted to monoglycerides, and monoglycerides are then converted to esters (biodiesel) and glycerol (by-product).

Microbial fuel cells

Microbial fuel cell (MFC) is another way of obtaining electricity from the hydrolysis and fermentation of algae in a single-step process. MFCs consist of an anode and cathode connected by a load (usually a resistor in laboratory studies). The anode contains mixed or pure cultures of microorganisms that are used to catalyze the decomposition of the organic matter into electrons and protons. Power is produced through the reduction of oxygen or another chemical at the cathode. A metal or some microorganism is usually used to catalyze oxygen reduction (Velasquez-Orta, Curtis, and Logan, 2009).

Velasquez-Orta, Curtis, and Logan evaluated the performance of MFCs using two different types of algae as substrates: *Chlorella vulgaris* (a microalgae) and *Ulva lactuca* (a macroalgae), as they have been utilized in other technologies for energy generation. Each has its particular organic-matter composition: *C. vulgaris* contains more than 50% protein and *U. lactuca* has around 60% carbohydrates. The substrate degradation and microbial composition in MFCs and in anaerobic reactors were compared.

MFCs using either microalgae or macroalgae produced relatively high power densities compared to the use of other substrates in this MFC configuration. Maximum power densities were best obtained either by using fixed resistances in the circuit for a sufficiently long time for the system to reach steady conditions, or by LSV using a slow scan rate (0.1 mV/s). *C. vulgaris* gave the highest energy generation per gram of substrate, while the macroalgae *U. lactuca* was degraded more efficiently in MFCs. The substrate composition and bioprocess had an effect on the final chemical oxygen demand (COD) removal obtained and the type of microbial communities present. Carbohydrates contained in *U. lactuca* were more completely degraded than proteins in *C. vulgaris*. Bioelectricity production using algae in MFCs is useful as a low-temperature method of power generation, but it needs to be further improved in order to make it competitive with alternative energy technologies (Velasquez-Orta, Curtis, and Logan, 2009).

Algae as Human Food

Hunger is a term which has three meanings (*Oxford English Dictionary* 1971):

1. "the uneasy or painful sensation caused by want of food; craving appetite. Also the exhausted condition caused by want of food"
2. "the want or scarcity of food in a country"
3. "a strong desire or craving"

Malnutrition "is a general term that indicates a lack of some or all nutritional elements necessary for human health." In general, there are two types of malnutrition. The first—considered most important, and the one usually referred to when hunger is discussed—is 'protein–energy malnutrition.' It is a lack of both calories (food energy) and protein. Protein–energy malnutrition is considered the most dangerous and lethal form of hunger. The second type is 'micronutrient malnutrition,' which is related to vitamin and mineral deficiency. This is not the type of malnutrition that is referred to when world hunger is discussed, though it is certainly very important ("2011 World Hunger," 2011).

Algae can be a solution for both types of malnutrition, as a source of protein, energy, vitamins, and minerals.

Algae and cyanobacteria as human food

Algae are major sources of food for many living creatures, including freshwater animals, seawater animals, domesticated animals, and some wild animals. Human consumption of algae emerged long ago, mainly among coastal people of east Asia and the Pacific such as the Chinese, Japanese, Koreans, Filipinos, and Hawaiians (Aaronson, 2000).

Some strains of algae are still in use for food today, including Chlorophyta, Phaeophyta, Rhodophyta, and cyanobacteria, either as direct food or as food additives, all over the globe. Using algae as food or as an ingredient in foods is more commonly found, however, in marine or coastal regions. People consuming algae vary widely in culture, geographic location, and background, from hunter-gatherers to industrial consumers (table 9.7). The way algae are consumed also varies, some being served raw and added to salads, others pickled, others fermented into relish. They are added to soups and dressings, and served as dessert. Some industrialized countries even produce alginates and carrageenans (algal products) from seaweeds as a substitute for other food or for variety in the diet (Aaronson, 2000).

Nutritional value of algae for human consumption

One of the most common species of algae, and the oldest in use as human food, is spirulina. Owing to its outstanding concentration of non-conventional nutrients, it has earned a reputation as a 'super-food,' and because of its high protein yield, scientists gave it the notable title of 'food of the future' (Henrikson, 1989).

In an attempt to alleviate protein deficiencies worldwide, intense research has been conducted in search of new alternative proteins. Some of this research involves single-cell protein extraction from microbial biomass such as yeast, bacteria, and microalgae. Due to technical and economic difficulties, and unlike other microbial biomass, extraction of protein alone from alga is of less interest to commercialists than the biomass as a whole. This is hardly a disadvantage,

Table 9.7. Use of algae in historical civilizations (Aaronson, 2000)

Country	Type	Used in
Vietnam	Azolla	Rice plantation
Coast of Peru (2500 BC) at Pampa, Playa Hermosa Concha, Gaviota, and Ancon	Kelp	Marine algae were used by the ancient Peruvians to supplement their diets
Valley of Mexico, Aztec times (13th–16th centuries)	*Spirulina maxima*, or *Spirulina geitleriai*	Dried and made into cakes, used as food
Chad, Africa (Kanembu people of Chad), 9th-century Kanem Empire	*Spirulina platensis*, species of cyanobacterium	Collected, dried into cakes called *dihe* or *die* to make broths for meals; made into a sauce
China, 260–220 BC AD 533–44 southern Fujian Province, AD 960–1279	*Ecklonia* seaweed *Porphyra* seaweed Mainly *Gloiopeltis furcata*	Food and medicine Food
Japan (10,500–300 BC) and the Yayoi period (200 BC to AD 200)	*Gelidium, Laminaria, Porphyra*, and *Undaria peterseneeniana*	Important food source
Greeks (70–19 BC) and Romans (65–8 BC)	*Algae vilior alga*	Emergency food for livestock
Iceland, AD 960	*Rhodymenia palmata*	Food

however, since microalgal biomass holds much nutritional value beyond protein alone. Besides its exceptional protein worth, algae has a "broad spectrum of other nutritious compounds, viz., peptides, carbohydrates, lipids, vitamins, pigments, minerals, and other valuable trace elements" (Becker, 2007), compared to the normal vegetables available to many people. Table 9.8 shows the nutritional constituents of different types of algae.

Table 9.8. General composition of different algae, as % of dry matter (Becker, 2007)

Alga	Protein	Carbohydrates	Lipids
Anabaena cylindrica	43–56	25–30	4–7
Aphanizomenon flosaquae	62	23	3
Chlamydomonas rheinhardii	48	17	21
Chlorella pyrenoidosa	57	26	2
Chlorella vulgaris	51–58	12–17	14–22
Dunaliella salina	57	32	6
Euglena gracilis	39–61	14–18	14–20
Porphyridium cruentum	28–39	40–57	9–14
Scenedesmus obliquus	50–56	10–17	12–14
Spirogyra sp.	6–20	33–64	11–21
Arthrospira maxima	60–71	13–16	6–7
Spirulina platensis	46–65	8–14	4–9
Synechococcus sp.	63	15	11

Table 9.9 suggests that an alga such as spirulina may be a perfect protein source, containing 65% protein, higher than any of the other common protein-rich dietaries such as fish, soybeans, dried milk, peanuts, eggs, grains, and whole milk.

Table 9.9. Protein % from different sources
(Henrikson, 1989)

Natural protein source	% protein
Spirulina	65
Animal and fish meat	15–25
Soybeans	35
Dried milk	35
Peanuts	25
Eggs	12
Grains	8–14
Whole milk	3

A short scientific description of the way the human body handles protein may be useful in explaining spirulina's valuable protein content. Two basic types of usable amino acids, the 'building blocks of proteins,' are required by the body, essential amino acids and non-essential amino acids. Since essential amino acids cannot be synthesized within the body, they need to be obtained from the diet. Only a protein that contains all the essential components of essential amino acids can be considered complete and able to meet the body's protein requirements. Spirulina is one protein source that contains all the essential amino acids. Table 9.10 shows the essential amino acid composition of spirulina.

Table 9.10. Spirulina's amino acid content (Henrikson, 1989)

Amino Acids	Per 10 gms	% total
Essential amino acids		
Isoleucine	350 mg	5.6
Leucine	540 mg	8.7
Lysine	290 mg	4.7
Methionine	140 mg	2.3
Phenylalanine	280 mg	4.5
Threonine	320 mg	5.2
Tryptophane	90 mg	1.5
Valine	400 mg	6.5
Non-essential amino acids		
Alanine	470 mg	7.6
Arginine	430 mg	6.9
Aspartic acid	610 mg	9.8
Cystine	60 mg	1
Glutamic acid	910 mg	14.6
Glycine	320 mg	5.2
Histidine	100 mg	1.6
Proline	270 mg	4.3
Serine	320 mg	5.2
Tyrosine	300 mg	4.8
Total amino acids	**6200 mg**	**100**

While spirulina's exceptional amino acid composition alone makes it outstanding food, its nutritional value is not limited to proteins. Spirulina contains remarkable amounts of vitamins, minerals, and nutrients, as shown in figure 9.8.

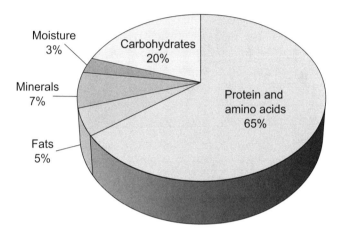

Figure 9.8. Spirulina's composition (Henrikson, 1989)

It has been shown that only 3 to 10gm of spirulina per day provides the right amounts of vitamins required for human health. Table 9.11 shows the vitamin content of spirulina, and table 9.12 shows the beta carotene content of spirulina as compared to other sources.

As table 9.12 shows, 10gm of spirulina contains 23,000 IU of beta carotene (14mg), a much higher amount than that found in carrots. Vitamin A is vital for maintaining healthy vision and its deficiency can lead to blindness. Beta carotene and vitamin A also reduce serum cholesterol and lower the risk of cancer.

Of all foods, spirulina also has one of the highest contents of vitamin B12, exceeding that found in beef liver, chlorella, and sea vegetables. Vitamin B12 is essential for the conversion of fats, proteins, and carbohydrates into energy. It is also important for keeping the immune system in good condition, maintaining healthy red blood cells, and thus avoiding heart diseases. Table 9.13 presents the mineral content of spirulina. Tables 9.14 and 9.15 compare the iron content of spirulina to other food types.

Table 9.11. Vitamin content of spirulina (Henrikson, 1989)

Vitamins	per 10 gm	US RDA	% RDA
Vitamin A (beta carotene)	23000 IU	5000 IU	460
Vitamin B1 (thiamine)	0.31 mg	1.5 mg	21
Vitamin B2 (riboflavin)	0.35 mg	1.7 mg	21
Vitamin B3 (niacin)	1.46 mg	20 mg	7
Vitamin B6 (pyridoxine)	80 mcg	2 mg	4
Vitamin B12	32 mcg	6 mcg	533
Vitamin C		60 mg	
Vitamin D		400 IU	
Vitamin E (a-tocopherol)	1 IU	30 IU	3
Folacin	1 mcg	400 mcg	0.25
Pantothenic acid	10 mcg	10 mg	1
Biotin	0.5 mcg		
Inositol	6.4 mg		

Spirulina is extremely rich in iron, richer than any other food source, not only in the amount of iron, but also in its quality; that is, the form of iron in spirulina is easily absorbed by the human body (60% better absorption than conventional iron supplements). Iron is important to keep red blood cells healthy and the immune system in good condition (Henrikson, 1989).

While spirulina is well known for its rich protein, iron, and vitamin nutritional quality, it is rarely mentioned as a calcium source. It is, however, rich in calcium, containing more calcium per gram than milk does. Calcium is necessary for the bones and neural transmissions to the muscles (Henrikson, 1989).

Spirulina also has high concentrations of magnesium, which is necessary for the absorption of calcium as well as the regulation of blood pressure (Henrikson, 1989). Last but not least, spirulina has some essential fatty acids like GLA that help regulate the formation of the vital hormones, prostaglandins. These hormones are responsible for regulating blood pressure, cholesterol, inflammation, and cell growth. Spirulina is one of very few sources of GLA, and is the richest, besides breast milk, evening primrose, blackcurrant, and borage seeds (Henrikson, 1989).

Table 9.12. Beta carotene of spirulina compared to other sources (Henrikson, 1989)

Food	Serving size	IU of beta carotene
Spirulina	1 heaping tsp. (10 gm)	23000
Papaya	½ medium	8867
Sweet potato	½ cup, cooked	8500
Collard greens	½ cup, cooked	7917
Carrots	½ cup, cooked	7250
Chard	½ cup, cooked	6042
Beet greens	½ cup, cooked	6042
Spinach	½ cup, cooked	6000
Cantaloupe	¼ medium	5667
Chlorella	50 tablets (10 gm)	5550
Broccoli	½ cup, cooked	3229
Butternut squash	½ cup, cooked	1333
Watermelon	1 cup	1173
Peach	1 large	1042
Apricot	1 medium	892

Table 9.13. Mineral content of spirulina (Henrikson, 1989)

Minerals	per 10 gm	US RDA	% US RDA
Calcium	100 mg	1000 mg	10
Iron	15 mg	18 mg	83
Zinc	300 mcg	15 mg	2
Phosphorus	90 mg	1000 mg	9
Magnesium	40 mg	400 mg	10
Copper	120 mcg	2 mcg	6
Iodine		150 mcg	
Sodium	60 mg	2–5 gr	1
Potassium	160 mg	6 gr	3
Manganese	500 mg	3 mg	17
Chromium	28 mcg	200 mcg	16
Germanium	6 mcg		
Selenium	2 mcg	100 mcg	2

Table 9.14. Iron content of food (Henrikson, 1989)

Food	mg Iron per 100 grams
Spirulina	150
Chlorella	130
Hazelnuts	8.1
Pistachios	6.7
Cashews	6.4
Beef liver (fried)	5.7
Whole-wheat bread	3.2
Pinto and white beans (boiled)	3
Spinach (raw)	2.7
Hamburger (cooked, lean)	2.7
Raisins, seeded (muscat)	2.6
Eggs	2
Lamb chops (broiled)	1.8
Soybean curd	1.8
Bean sprouts	1.6

Table 9.16 summarizes the nutritional value of spirulina and its consequent potential benefits to the human body.

Today, because of the great nutritional value of photoautotrophic microalgae, advanced technologies are used to produce commercial amounts of about 10,000 tons per year (Becker, 2007). Biomass from microalgae is dried and marketed in the human health-food market in the form of powders or pressed in the form of tablets (Demirbas and Demirbas, 2010). The most common types of algae processed in mass production include chlorophycea, *Chlorella sp.*, and *Scenedesmus obliquus*, and the cyanobacteria *Spirulina sp.* and *Athrospira sp.* (Becker, 2007). Table 9.17 shows how the different forms of spirulina are consumed.

Algae potential as a global human food

As with all food products released in the market, algae had to undergo numerous toxicological tests to prove that they are safe for human consumption and free of harmful materials. To date, algae have shown no

Table 9.15. Iron sources and serving sizes (Henrikson, 1989)

Food	Serving size	mg Iron
Spirulina	1 tbsp. (10 grams)	15
Chlorella	1 tbsp. (10 grams)	13
Chicken liver, cooked	3 ounces	7.2
Crab pieces, steamed	½ cup	6
Beef liver, fried	½ cup	5.3
Soybeans, boiled	½ cup	4.4
Blackstrap molasses	1 tbsp.	3.2
Spinach, cooked	½ cup	3.2
Beef, sirloin, broiled	3 ounces	2.9
Potato, baked	1	2.8
Scallops, steamed	3 ounces	2.5
Pistachios, dried	¼ cup	2.2
Broccoli, cooked	1 spear	2.1
Cashews, dry-roasted	¼ cup	2.1
Lima beans, boiled	½ cup	2.1
Turkey, dark meat	3 ounces	2
Spinach, raw, chopped	½ cup	0.8

toxicity and no side effects resulting from their use. More specifically, none of the algae proposed as human food have imposed any negative impacts on humans, in any of the short- or long-term experiments. Human algae consumption is therefore regarded as harmless as long as the algae are properly processed (Becker, 2007).

Not only is algal biomass safe to consume, it also contains a potential protein source that is considered at least equal to other plant proteins, and in most cases superior to them, as discussed in the previous section (Becker, 2007).

In spite of their nutritional value, and the fact that there are no restrictions on their use, dried microalgae are still not widely accepted as a source of protein. The main reasons go back to the consistency of the dried biomass, its color, and its smell. Most attempts to add dried algae to baked items have resulted in a considerable change in taste. In

Table 9.16. Essential nutrients in spirulina (Henrikson, 1989)

Essential nutrients	Content in spirulina	Importance
Vitamin A	Very high: higher than carrots	Maintain healthy vision; deficiency could lead to blindness
Vitamin B12	Very high: richer in B12 than common whole grains, beef liver, fruits, and vegetables	Conversion of fats, proteins, and carbohydrates into energy Enhance immune system Healthy red blood cells, thus help avoid heart diseases
Beta carotene	Very high: much higher than carrots	Reduce serum cholesterol Lower the risk of cancer
Fatty acids	Very high: richest of the very few sources of GLA	Formation of prostaglandins that regulate blood pressure, cholesterol, inflammation, and cell growth
Iron	Very high	Red blood cells Healthy immune system
Calcium	More calcium per gram than milk	Bones Neural transmission to the muscles

some parts of the world, microalgae face problems because of socioeconomic barriers and reluctance to accept new food ingredients. Furthermore, the production costs for safe processing of microalgae are still high, thus restricting its competitive value with other protein sources (Becker, 2007).

Algae Potential as Animal Feed

The production of animal feed is a major source of income worldwide, and the feed itself is an indirect source of the food on which human life is based. A survey of international markets showed that the world feed production is highly monopolized. Annual output is around 614 million

Table 9.17. Forms of spirulina available for consumption

Spirulina form	Method of use/dosage
	Soups, salads, pasta, breads, and taboula, or added to fruit juice in a blender
	Sprinkled on foods such as salads or cereals or eaten directly as a snack
	Swallowed directly. Tablets usually contain 500g, capsules contain 400g. Most people take 20 tablets a day, equivalent to 1 heaping teaspoon of powder.

tons, 70% of which is produced by only four countries and 20% produced by another 44 countries (Algae for Feed, 2010). The European Union employs 110,000 people in the field of feed production, which helps to enrich the economies of the countries involved. Every year the feed industry provides food for 6 billion broiler chickens, 370 million laying hens, 260 million pigs, 90 million bovines, and 100 million sheep and goats. It provides humans with 130 million hectoliters of safe milk, 45 million tons of safe meat, and 6 million tons of safe eggs (Algae for Feed, 2010). This large productivity means that the feed industry affects our daily lives considerably, and finding cheap, cost-effective, high-nutrient feed alternatives means providing for a better living and a better economy. The survival and growth of animals depends on their health. If an element such as algae, with high nutrient value and no chemical additives, can be used as feed, the economy in general and the feed industry in particular would benefit (Sweetman, 2009; Pulz and Gross, 2004). The aquacultural feed industry already utilizes a large percentage of algal production (Becker, 2004).

Although algae have had a long history in sustaining aquatic life and fish, using microalgae in animal feed is a relatively new concept. Research has recently shown, however, that algae's nutritional values are not restricted to humans and fish alone, but offer numerous health benefits for various animals and birds.

Spirulina, added to animal feed, was found to have direct, noticeable health effects on animals, birds, and fish, including improved growth rates, decrease in mortality, and increase in immunity, as summarized in tables 9.18 and 9.19 (Fannin, 2012).

Tests on mice and other species also revealed many other positive effects when a percentage of proteins in animal diets is replaced by spirulina or spirulina extracts. They include reduction of cholesterol and anticancer effects in mice and humans, as shown in table 9.20 (Fannin, 2012).

It is estimated that around 30% of the current algal production is used for feeding animals (Becker, 2007). In addition to contributing to the general health of animals and prolonging their lives, algae also have major commercial benefits.

Laminaria, ascophyllum, and fucus seaweeds can be used as animal fodder, fed directly to cattle after being cultivated, harvested, dried, and ground (Faheed and Abdel Fattah, 2008). Algae species like pelvetia, when added to fodder, increase the milk yield by 10%. Using algae in poultry feed also provides a high commercial return.

Microalgae have also found their way into aquaculture, where demand for it is progressively growing.

Algae Use in Farming

Modern agriculture has focused on obtaining high short-term yields, excessively using seed hybrids, considerable amounts of inorganic fertilizer and pesticides, and abusing water and energy resources, while paying no attention to the unseen consequences of depleting nonrenewable fossil fuels and polluting soil and water in the process (Henrikson, 1989).

Egypt presents a typical scenario of agricultural land use, where for years farming followed the universal shallow techniques. The heavy reliance on chemical fertilizers continually damages both the environment and the quality of crops. The fertility of the soil is progressively disappearing because of soil erosion, diminishing of natural soil nutrients, and buildup of salts, toxic elements, and waterlogging (Faheed and Abdel Fattah, 2008).

Despite the fact that a simple solution may be readily available, "complex and sophisticated technology seems to be taking over, blinding the world to large volume, low-tech solutions that any farmer can understand and implement" (Ibrisimovic, 2008).

Spirulina was suggested by the FAO in 1981 as a possible replacement for chemical fertilizers and soil structure rebuilder. In India spirulina is grown in earthen ponds, dried by evaporation, and sold to rice farmers. It can be a nitrogen source at one-third the cost of chemical fertilizer, and it increases the annual yield of rice by 22%. Spirulina was also able to increase tomato yield (Zeenat and Sharma, 1990). Research has been conducted in Egypt using algae to purify soil from heavy metals and clean up oil spills by feeding on them (Sawahel, 2004).

Biofertilizers, according to Faheed and Abdel Fattah (2008), may provide the only future solution to soil nutrient deficiency and sustainable agriculture. Algae may in fact introduce the world to what may be called an "Agricultural Revolution" (Ibrisimovic, 2008), where any algae that are unsuitable for other applications, like food and fuel, may be readily processed to produce excellent quality soil conditioning or fertilizer. Biofertilizers mainly contain microorganisms that improve soil fertility through their nitrogen-fixing, phosphate-solubilizing, and growth-promoting properties. Some of these include Azotobacter, Azospirillium, blue-green algae, Azolla, P-solubilizing microorganisms, mycorrhizae, and Sinorhizobium (Faheed and Abdel Fattah, 2008).

Dry green algae, for example, "contains high percentage of macronutrients, considerable amount of micronutrients and amino acids and are simply cultivated on sewage and brackish water for use instead of chemical fertilizers" (Faheed and Abdel Fattah, 2008). A study, reported in Faheed and Abdel Fattah's article, was carried out in Egypt to examine the effect of fresh and dry green microalgae (*Chlorella vulgaris*) on the output yield of the soil, crop quality, and environmental improvement. The study focused on the effect of fertilization treatment of lettuce plants by microalgae. The results showed that fresh and dry green microalgae improved the growth parameters and the pigment content. The fresh and dry weights were also found to increase with the algae treatment as a result of increased nutrient levels in the soil.

Another study was carried out in India testing the effect of blue-green algae on rice cultivation. Despite the soaking environment of rice fields, blue-green algae were found to have the ability to survive and positively affect the crop. Blue-green algae, cyanobacteria in particular,

Table 9.18. Effects of spirulina on growth and survival (Belay et al., 1996)

Subject
Big mouth buffalo *(Ictiobus cyprinellus)*
Nile tilapia *(Tilapia nilotica)*
Milkfish fry *(Chanos chanos)*
Silver carp *(Hypothalamichthys molitrix)*
Cultured striped jack *(Pseudocaranx dentex)*
Yellow tail *(Seriola quinqueradiata)*
Masou salmon *(Oncorhynchus masou)*
Sea eels *(Anguilla japonica)*
Red sea bream *(Pagrus major)*
Silver sea bream *(Rhabdosargus sarba)*
Giant freshwater prawn *(Macrobrachium rosenbergii)*
Chicken
Quail
Turkey poults

Result
Improved growth rate at 29g dwt/kg-l body wt
Spirulina supplemented with methionine successfully replaced fish meal
Substitution of formulated diet with up to 50% spirulina was possible without adverse effect
10% spirulina in basal diet improved specific growth rates and live weight 30% of protein of formulated feed replaced by spirulina with improved growth
5% spirulina in feed resulted in good growth rate and feed conversion efficiency; 10% spirulina suppressed growth slightly
1–10% spirulina supplementation resulted in improved growth rate (1.5 times control), survival rate, and feed efficiency
Improved growth rate, survival rate, feed conversion efficiency, resistance against bacterial infection
Improved growth rate, survival, feed efficiency
5% spirulina in feed showed significant increase in growth rate, feed conversion efficiency
Up to 50% spirulina substitution gave comparable growth rate to control without adverse effect
5–10% spirulina significantly improved growth, survival, and feed utilization
1.5–12% spirulina in diet of male broiler chicks can substitute for other protein source with good growth rate and feed efficiency
5–10% spirulina in feed of growing chickens was found to improve growth in chickens and laying hens
1.2–10% spirulina in feed improved fertility, hatchability, and egg production
0.2% spirulina reduced the death rate to as low as 1/20 that of control
1,000–10,000 mg/kg-I spirulina produced significantly higher growth rate and lower mortality rate
Higher growth rate and lower mortality rate

Table 9.19. Immunomodulatory effect of spirulina (Belay et al., 1996)

Subject	Result
Channel catfish	Feeding fish spirulina-enriched Artemia (bioencapsulation) resulted in a 3-fold increase in splenic macrophage number, and increased activation of macrophytes compared to Artemia feed alone
Chicken	In vitro treatment of macrophages with a water extract of spirulina resulted in enhanced macrophage activation (phagocytosis)
	In vivo feeding to K. strain Leghorns (up to 10,000 mg/kg) showed larger thymi, higher natural killer cell (NK) activity and CBH response; number of phagocytic macrophages increased and number of SRBC/phagocytic macrophages also greater in spirulina-fed group
	Significantly higher bacterial clearing rates observed with spirulina supplementation compared to control diet
Turkey	Improved lymphoid organ development: splenic bursal and thymic weights were higher in the spirulina-fed group

Table 9.20. Animal testing of spirulina and consequent results (Belay et al., 1996)

Application	Subject
Cholesterol reduction	Mice
	Humans
Protection against nephrotoxicity	Mice
Anti-cancer effects	Mice
	Hamsters
	Humans
Radiation protection	Mice
Marrowcytes/stem-cell regeneration	Mice
Antiviral effects	Mice
Immunomodulatory effects	Mice
	Cats

proved capable of enriching the soil with cheap nitrogen, increasing the crop by improving the soil's fertility. Under controlled conditions, blue-green algae provided the soil with 25–30kg of nitrogen per hectare per season, raising the yield by 10–15%. Other advantages include "stabilization of soil, add organic matter, release growth promoting substances, improve soil physico-chemical properties and solubilize the insoluble phosphates" (Mishra and Pabbi, 2004).

Blue-green algae also increase microbial activity and discourage weeds by stopping nutrients and light from reaching them. Blue-green algae were found to enhance general soil properties, ensure an uninterrupted nutrient supply, improve the water-holding capacity of the soil, and enhance aggregation.

Producing fertilizers from algae will reduce farmers' usage of costly chemical fertilizers that pollute the environment, while giving them the option of growing fish with algae in barren land as well as producing crops normally on fertile areas of their farms.

Most important of all, the technology is affordable, quite simple, and can be used by most farmers easily.

Use of Algae in Water Treatment
Industrial wastes
Waste water containing organic materials and heavy metals has increased in volume with ever-increasing consumer demands. Such wastes, however, cannot be simply discharged into water resources before treatment. Despite the presence of many physico-chemical ways to treat water containing such pollutants, most conventional methods are expensive and labor-intensive, and do not meet required standards when pollutants are present in high concentrations (Mehta and Gaur, 2005).

Cyanobacteria and algae are non-conventional wastewater treatment elements that assist in the removal of nutrients like phosphorus and nitrogen from contaminated water, while at the same time producing biomass (Mehta and Gaur, 2005). Cyanobacteria and microalgae also provide the water with oxygen, promoting the bacterial action of oxidation of organic material. Anabaena and ellipsosporum, for example, are able to break down organic contaminants such as antibiotic wastes of penicillin and ampicillin that may have annoying odors and adverse environmental effects, by using them as sources of nitrogen. They are also able to grow in a high phenol concentration, as well as remove phenol in amounts of 38 mg/L in a 7-day retention period. The alkaline environment that results from spirulina is then used to precipitate the heavy metals from the effluents (Mehta and Gaur, 2005). Table 9.21 presents a comparison of the heavy metal uptake capacities of various macrophytes.

Cyanobacteria are capable of clearing the water of heavy metals, even if more than one metal is present. This ability is due to several functional groups—including carboxyl, amino, sulfate, and hydroxyl—that house the adsorbed metals (Kumar, Oommen, and Kumar, 2009).

Biosorption consists mainly of the attaching of metals onto cell surfaces and intracellular ligands. This is usually carried out in two steps. Metals are first adsorbed on the cells' surface. This is a fast process but, since it is only physical, it has a very short effect and can easily be reversed. The second process is a chemical metabolic reaction that includes a slow uptake of metals into the cytoplasm, where they are either stored or converted. The first step is usually responsible for 80% of the heavy metal removal, and so the better the adsorption is carried out, the more efficient the process is.

Several factors determine how well biosorption is carried out. Some of these are the pH, the temperature, the presence of ions (positive and negative), the amount of nutrients present, the lighting available, and the efficiency of the metabolism of the microorganisms.

The adsorption usually happens at the cell wall, which contains several functional groups such as hydroxyl (-OH), phosphoryl ($-PO_3O_2$), amino ($-NH_2$), and carboxyl (-COOH), that are responsible for the wall's negative charge (Ahalya, Ramachandra, and Kanamadi, 2003)

The main processes that contribute to biosorption are ion exchange, complexation, electrostatic attraction, and microprecipitation, depending on the algal species involved and the surrounding conditions. Ion exchange is considered the most common process in heavy metal biosorption and is generally carried out more effectively when the biomass is treated with agents such as $CaCl_2$ or NaOH that raise their binding capacity (Ahalya, Ramachandra, and Kanamadi, 2003).

Organic wastes and algae oxidation ponds

After being treated by primary clarification, which removes the settleable solids, sewage is passed through an oxidation pond containing algae, as shown in figure 9.9. An oxidation pond is a manmade basin that may be 3 to 5 feet deep where the water is kept for about a month. Microorganisms absorb oxygen from the surrounding environment and utilize it to convert the carbon waste in the organics to carbon dioxide

Table 9.21. Comparison of heavy metal uptake capacities (mg/g) of various macrophytes (Sharma et al., 2011)

Adsorbent	Lead	Zinc	Copper
M. spicatum	46.69	15.59	10.37
M. spicatum (Wang et al., 1996)	55.6	13.5	12.9
P. lucens	141	32.4	40.8
S. herzegoi	—	18.1	19.7
E. crassipes	—	19.2	23.1

gas. This prepares the environment for algae growth. Algae utilize the carbon dioxide that was produced by the aerobic bacteria, along with light from the sun, to convert it to algal cells and oxygen, as shown in figure 9.10. Since algae usually produce extra oxygen in excess of their own need, some of this oxygen may be transferred to the inlet of the pond where it is most needed (Sharma et al., 2011).

Sewage also contains considerable amounts of nitrogen, present as ammonia and organic nitrogen, a nutrient that is needed for algae growth. The bacteria–algae interaction can be used to retrieve waste nitrogen through the formation of algal cell material. The algae have the capability of using both the ammonia contained in the sewage and the ammonia and nitrate resulting from bacterial processes (Sharma et al., 2011).

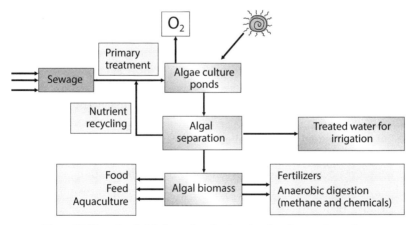

Figure 9.9. Flow of materials in an algae treatment system for organic wastes (Sharma et al., 2011)

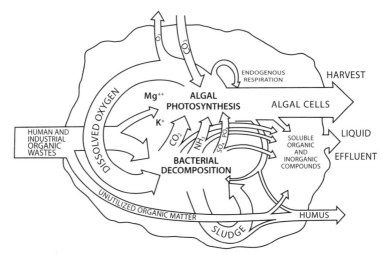

Figure 9.10. Algae–bacterial interaction (Gottas, Oswald, and Ludwig, 1954)

Algae for Biological Fixation

Large amounts of carbon dioxide are emitted into the atmosphere from energy production. Elimination of CO_2 has been done through various physico-chemical methods and separation technologies, such as dry and wet absorption, adsorption, and membrane. The collected CO_2 has been utilized in plastics, paint, solvents, cleaning compounds, construction materials, and packaging (Zhang et al., 2010). Still, considerable amounts of CO_2 are emitted into the atmosphere.

One of the options investigated for CO_2 capture is biological fixation using microalgae photosynthesis. This technique depends on the properties of microalgae, including efficient photosynthesis superior to terrestrial plants, fast proliferation rates, wide tolerance to extreme environments, and potential for intensive cultures (Zhang et al., 2010). Carbon dioxide emissions are reduced because the carbon is fixed by microalgae and converted to carbohydrates, lipids, and proteins. Hence, algal biomass can produce energy, chemicals, and food.

One way to accomplish this is to direct the flue gases of conventional fossil-derived power plants or any industrial combustion process to desulfurization units to separate the CO_2 and NO_x emissions. These gases are then directed to algae cultivation ponds. The oil extracted from the resulting algae broth is refined, to produce biodiesel and by-products such as glycerol, as shown in figure 9.11.

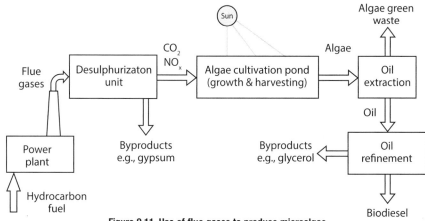

Figure 9.11. Use of flue gases to produce microalgae

It is more economically feasible to capture CO_2 from the discharge gases of heavy industries containing 10–20% CO_2 levels than from the atmosphere, which contains 0.036% CO_2. *Chlorella* was found to be better than *Isochrysis sp.* and *Amphidinium carterae* for eliminating high levels of CO_2. The CO_2 fixation rate of *Chlorella* cultivation with 15% CO_2 was 2979 mg/L per week. This implies that *Chlorella* has great potential advantages in microalgal biomass energy (Zhang et al., 2010).

Other Uses of Algae
Algae with powerful anti-cancer properties
Dunaliella salina is a green, single-celled alga that grows in oceans and salt lakes. It is full of beta carotene, which has properties that counteract cancer. It reduces the harmful radicals in the human body that, if not decreased, can damage the deoxyribonucleic acid (DNA), increasing the risk of cancer. Algal beta carotene is more powerful than the type found in vegetables and fruits (John, 2001).

Researchers were surprised to find that *Dunaliella* has certain properties that allow it to adapt to the surrounding environment, whatever it might be. For example, it produces antioxidants as a response to the harmful free radicals and other poisonous toxins it encounters. In an experiment performed at the chemistry department at El Carmen University in Huelva, Spain, scientists exposed the *Dunaliella* algae to a very intense damaging light that encourages the production of free radicals. This damage is similar to that caused by exposure to the sun, which results in wrinkles. However, they found that this *Dunaliella* started to produce antioxidant chemicals called carotenoids, including beta carotene (John, 2001). In another experiment, the plant was transferred from the sea to the laboratory. As a result, they found that the plant produced carotenoids, a self-defense mechanism to protect it against oxygen starvation. However, the resulting carotenoids (including beta carotene) are considered to be of a very high quality. What is interesting here is that the beta carotene produced by the *Dunaliella* has much stronger properties, and can absorb higher amounts of ultraviolet radiation coming from the sun compared to the amount that ordinary beta carotene is able to absorb (John, 2001).

Use of algae in cosmetics

The use of algae is not limited to feed, food, and fuel; algae extracts have also found valuable uses as cosmetics. A cosmetic product with a label including the words "marine extract," "extract of alga," or "macroalga extract" usually implies that the ingredients include a hydrocolloid extract from macroalgae. Some forms of macroalgae in cosmetic uses include sachets of milled macroalgae with essential oil sometimes added to bathwater, bath salts with macroalgae, creams, face masks, shampoos, body gels, bath salts, and even do-it-yourself body wrap kits. In thalassotherapy, which has recently come into vogue, macroalgae pastes are produced by cold-grinding or freeze-crushing, then spread on the body and exposed to heat using infrared radiation.

Sometimes the algae treatment involves wrapping the body after spreading it with microalgae paste. Such treatment claims to remove cellulite and relieve rheumatic pain. Algae massage creams are also said to restore elasticity to the skin and help retain skin moisture.

Questions

1. Why is fuel derived from biomass considered more sustainable than fossil fuel?
2. List the advantages of producing biofuel from algae.
3. Explain why algae are recommended in harsh environmental conditions.
4. Discuss the uses of different algae as a source of food in ancient civilizations.
5. Describe briefly the two main groups of algae.

6. Discuss the four generations of biofuels in terms of their feed stock.
7. List the microalgae biodiesel value chain stages.
8. Discuss the three types of industrial reactors used for algal culture.
9. Explain the importance of spirulina as a food source.
10. List some of the uses of microalgae, and discuss the use of microalgae in waste-water treatment.

References

Aaronson, S. 2000. "Algae." In *Cambridge World History of Food*, chapter II.C.1. DOI: 10.1017/CHOL9780521402149.027

Ahalya, N., T.V. Ramachandra, and R.D. Kanamadi. 2003. "Biosorption of Heavy Metals." *Research Journal of Chemistry and Environment* 7, no. 4. wgbis.ces.iisc.ernet.in/energy/water/paper/biosorption/biosorption.htm

Algae for Feed. 2010. "Why Microalgae for Feeds." www.algae4feed.org /brief/why-microalgae-for-feeds/26

Becker, E.W. 2007. "Micro-algae as a Source of Protein." *Biotechnology Advances* 25, no. 2: 207–10.

———. 2004. "Microalgae in Human and Animal Nutrition." In *Handbook of Microalgae Culture: Biotechnology and Applied Phycology*, edited by A. Richmond. Oxford: Blackwell Science.

Belay, Amha, Kato Toshimitsu, and Ota Yoshimichi. 1996. "Spirulina (Arthrospira): Potential Application as an Animal Feed Supplement." *Journal of Applied Phycology* 8, no. 4–5: 303–11.

Chisti, Y. 2008. "Biodiesel from Microalgae Beats Bioethanol." *Trends in Biotechnology* 26, no. 3: 126–31.

———. 2007. "Biodiesel from Microalgae." *Biotechnology Advances* 25, no. 3: 294–306.

Demirbas, A., and M.F. Demirbas. 2010. *Algae Energy: Algae as a New Source of Biodiesel*. London: Springer-Verlag.

Edwards, M. 2009. *Green Solar Gardens: Algae's Promise to End Hunger; Solve Hunger Globally with SAFE Production—Sustainable and Affordable Food and Energy Grown Locally*. Tempe, AZ: CreateSpace.

Faheed, F.A., and Z. Abdel Fattah. 2008. "Effect of *Chlorella vulgaris* as Biofertilizer on Growth: Parameters and Metabolic Aspects of Lettuce Plant." *Journal of Agriculture and Social Sciences* 4, no. 4: 165–69.

Fannin, B. 2012. "Evaluating Algae Co-products as Potential Sources of Live-stock Feed: Algae for Fuel." *Texas A&M AgriLife.* agrilife.org/algaeforfuel/news/evaluatingcoproducts/

Gottas, H.B., W.J. Oswald, and H.F. Ludwig. 1954. "Photosynthetic Reclamation of Organic Wastes." *Scientific Monthly* 79, no. 6: 368–78.

Henrikson, R. 1989. *Earth Food Spirulina: How This Remarkable Blue-green Algae Can Transform Your Health and Our Planet.* Laguna Beach, CA: Ronore Enterprises.

Ibrisimovic, D. 2008. "Algal Farming: A New Agricultural Revolution?" Sciencealert, 29 April. www.sciencealert.com.au/opinions/20083004-17248-2.html

John, M. 2001. "Dunaliella Salina: The Algae with Powerful Anti-cancer Properties." www.thehealthierlife.co.uk/natural-health-articles/cancer/dunaliella-salina-algae-anti-cancer-properties-00119.html

Kumar, J.I.N., C. Oommen, and R.N. Kumar. 2009. "Biosorption of Heavy Metals from Aqueous Solution by Green Marine Macroalgae from Okha Port, Gulf of Kutch, India." *American-Eurasian Journal of Agricultural and Environmental Science* 6, no. 3: 317–23.

Marsh, G. 2009. "Small Wonders: Biomass from Algae." *Renewable Energy Focus* 9, no. 7: 74–76, 78.

Mata, T.M., A.A. Martins, and N.S. Caetano. 2010. "Microalgae for Biodiesel Production and Other Applications: A Review." *Renewable and Sustainable Energy Reviews* 14, no. 1: 217–32.

Mehta, S.K., and J.P. Gaur. 2005. "Use of Algae for Removing Heavy Metal Ions from Wastewater: Progress and Prospects." *Critical Reviews in Biotechnology.* DOI: 10.1080/07388550500248571

Mishra, U., and S. Pabbi. 2004. "Cyanobacteria: A Potential Biofertilizer for Rice." *Resonance,* June.

Pulz, O., and W. Gross. 2004. "Valuable Products from Biotechnology of Microalgae." *Applied Microbiology and Biotechnology* 65, no. 5: 635–48.

Roosa, S.A. 2010. *Sustainable Development Handbook.* 2nd ed. Lilburn, GA: Fairmont Press.

Sawahel, W. 2004. "Egyptian Algae Can Clean Up Oil and Soils." www.scidev.net/en/news/egyptian-algae-can-clean-up-oil-and-soils.html

Sharma, N.K., S.P. Tiwari, K. Tripathi, and A.K. Rai. 2011. "Sustainability and Cyanobacteria (Blue-green Algae): Facts and Challenges." *Journal of Applied Phycology* 23, no. 6: 1059–81. DOI 10.1007/s10811-010-9626-3

Sweetman, E. 2009. *Microalgae: Its Applications and Potential. International Aqua Feed.* Perendale Publishers.

"2011 World Hunger and Poverty Facts and Statistics." 2011. www.world-hunger.org/articles/Learn/world hunger facts 2002.htm

United Nations. 1987. "Report of the World Commission on Environment and Development." General Assembly Resolution 42/187, 11 December 1987.

Velasquez-Orta, S.B., T.P. Curtis, and B.E. Logan. 2009. "Energy from Algae Using Microbial Fuel Cells." *Wiley InterScience* 103, no. 6: 1068–76.

Zeenat, Rizvi, and V.K. Sharma. 1990. "Synergistic Effect of EB and Dap on Tomato Yield." *Science and Culture* 56: 129–30.

Zhang, Y., B. Zhao, K. Xiong, and Z. Zhang. 2010. "CO_2 Emission Reduction from Power Plant Flue Gas by Micro-algae: A Preliminary Study." Power and Energy Engineering Conference (APPEEC). Chengdu.

CHAPTER 10

Science and Technology Parks for Sustainable Development

Salah M. El-Haggar

Introduction

The idea of science and technology parks (STPs) arose during the 1940s and the 1950s when Frederick Terman, the dean of engineering at Stanford University, along with others, encouraged the faculty and graduates to establish their own high-tech firms surrounding the university. This is when the Science and Technology Park, now known as Silicon Valley, started to develop in San Francisco Bay. At first it was only a home for the electronics industry; then experiments and innovations in the fields of television, radio, and military electronics began.

The Silicon Valley Science and Technology Park, sometimes called Stanford University Science and Technology Park, transformed the region from one of the poorest in the United States into a global center of excellence in technology, innovation, finance, education, and research. An Australian government report stated that Stanford University's Industrial Park has not only enhanced the high reputation of that university, but also underpinned the high-technology industrial growth in Silicon Valley (UNESCO, 2012).

Silicon Valley was initially developed to meet the needs of entrepreneurial-minded academics. It was followed by Sophia Antipolis (France) in Europe in the 1960s and Tsukuba Science City (Japan) in Asia in the early 1970s (UNESCO, 2012) These science parks are the oldest and best known in the world.

In the United Kingdom, the science and technology movement started in 1971 with the establishment of science parks at Heriot-Watt University in Scotland and at Cambridge University. Today, there are over 400 science and technology parks worldwide and their number is still growing. The United States tops the list with more than 150 science parks. Japan follows with 111. China now has about 100. These numbers are expected to grow in the next few decades (UNESCO, 2012).

During the last few years, science and technology parks have been acting as a catalyst for regional economic development, promoting new technology-based companies and bridging the knowledge gap between academic research and industry. Science and technology parks thus contribute greatly to the 'innovation culture' and to boosting the competitive power of companies and institutions (Moudi and Hajihosseini, 2011). This kind of link between industry and universities was not generally encouraged until the 1970s, when the transfer of knowledge from universities to industrial companies became more frequent and the concept of science and technology parks was recognized.

There is no uniformly accepted definition of 'science park' or 'science and technology park,' and there are several other terms used to describe similar development, such as 'research park,' 'technology park,' 'business park,' 'innovation center,' 'technology valley,' 'incubator,' and so on (Monck et al., 1988).

The United Kingdom Science Parks Association (UKSPA) defines a science park as a business support and technology transfer initiative that: (1) encourages and supports the start-up and incubation of innovation-led, high-growth, knowledge-based businesses; (2) provides an environment where larger and international businesses can develop specific and close interactions with a particular center of knowledge creation for their mutual benefit; and (3) has formal and operational links with centers of knowledge creation such as universities, higher education institutes, and research organizations (UKSPA, 2013).

The Association of Universities and Research Parks (AURP) states that a university research park is a property-based venture, which plans property designed for research and commercialization; creates partnerships with universities and research institutions; encourages the growth of new companies; translates technology; and drives technology-led economic development (AURP, 2013).

The International Association of Science Parks (IASP) defines a park as "an organization managed by specialized professionals, whose main aim is to increase the wealth of its community by promoting the culture of innovation and the competitiveness of its associated businesses and knowledge-based institutions" (IASP, 2006).

To enable all of these goals to be met, an STP manages the flow of knowledge and technology among research and academic institutions and industry, encourages the establishment and growth of innovation-based companies, and provides other value-added services, such as high-quality work space and facilities.

The most common definition of STP is the one given by the IASP because it is the broadest, embracing all the existing experience of STP initiatives: "science and technology parks are sets of buildings where universities, industrial organizations and research institutes meet, cooperate and work together for the purpose of economic development using innovative ideas in technology, management, marketing, financing, promotion of products, etc." (Alberto, Perez-Canto, and Landoni, 2010).

Science and technology parks play an important role in technological capacity building, bringing together in the same location facilities that include manufacturing, training, technology and business incubation, and financing institutions. The science and technology environment encourages competitiveness in the field of scientific research in order to meet human demands. The idea of the STP is the promotion of inter-actions among science, technology, innovation, and economic development via its university–industry–government partnership. STPs are partnerships between industries, universities (and/or research insti-tutes), and governments. It is important to maintain strong links among the partners in order to avoid 'system failure.' System failure occurs when one of the links is (1) absent, (2) ineffective, or (3) inappropriate. In other words, industries, research communities, universities, and gov-ernments need an integrated mechanism that will transfer the results at the end to the industrial component.

Thus, STPs are a means for achieving economic and industrial devel-opment on a scientific base. The history of STPs shows the importance of the availability of universities and research institutes as assets. The link between these assets and the industry is essential for the success of STPs. The development of STPs faces particular challenges in devel-oping countries, where some of the required factors may be missing. Basic changes in the systems of these countries are needed to overcome these challenges.

The science and technology park can thus be defined as an organization to promote a culture of innovation, creativity, and competitiveness in the knowledge-based companies and institutions that are associated with the park, in order to increase the wealth of the community. STPs encourage interaction among science, technology, innovation, and economic development through the cooperation of universities, research centers, industries, and government.

The number of STPs registered with IASP, by region or continent, are: 11 in Africa, 215 in Western Europe, 10 in Eastern Europe, 38 in the Middle East, 37 in Asia, 38 in North America, and 6 in South America (UNESCO, 2012).

Objectives, Benefits, and Challenges of STPs

A science park has three main objectives. The first is to act as a catalyst for regional economic development. The second is to facilitate the creation and development of new technology-based companies and knowledge transfer from university to companies. The third is to provide the infrastructure of technical, logistic, and administrative support that a young firm needs in the process of struggling to gain a foothold in a competitive market.

In many cases, STPs are expected to generate benefits that go beyond regional development and job creation. Indeed, to the extent that STPs are effective, they have the potential to shift the terms of global competitiveness, not least in leading technological sectors. Therefore, they usually have educational and economic targets over and above the other objectives previously discussed. They are expected to increase competitiveness in research and scientific development in order to meet human demands, leading to higher achievements in the field of research, development, and innovations. STPs provide technical and scientific degrees as well as job opportunities after graduation. Research and innovations are dedicated to industrial development, which improves the well-being of society and raises the standard of living. STPs play a significant role in knowledge and resource management in the following ways (University of Pune, 2012):

- Bridge the knowledge and technology gap between universities and companies
- Create communication links between companies, entrepreneurs, and technicians, and allow knowledge and technology to flow between universities, research institutions, firms, and markets, as

well as provide consultancy services in the area of institutional and infrastructure development

- Promote innovation, creativity, and quality, and provide value-added services, including quality spaces and facilities
- Identify the different stakeholders, including companies, research institutions, entrepreneurs, and 'knowledge workers'
- Encourage the growth of small and medium-size companies and the initiation of new businesses by providing the necessary support
- Link thousands of companies and research institutes and create a global network to facilitate the internationalization of resident companies and the international integration of innovative firms
- Commercialize resources and clean technologies
- Transform ideas into technologies to serve the market and promote innovation on a regional scale
- Establish synergies between universities and industry
- Create new employment opportunities and knowledge-based jobs.
- Train workers, transfer technology, and serve entrepreneurs
- Create short-, medium-, and long-term strategic alliances and implement government policies
- Encourage innovative and technology-based entrepreneurial development
- Add value to the region in which it is located, by enhancing the quality of life and attracting talent
- Manage innovations, intellectual property, industry-sponsored research, and the development of entrepreneurship
- Boost economic growth and establish attractive working conditions for knowledge workers

The United Nations Educational, Scientific, and Cultural Organization (UNESCO) has pointed out the challenges that face the development of science and technology parks in developing countries. These include the availability of infrastructure such as land and utilities, funding, skilled manpower, lack of awareness about the concept of STPs and therefore lack of support, and a lack of sufficient legal and policy framework. UNESCO explains that, before establishing an STP in a developing country, several factors should be studied: To what extent is the economy ready to grow in the areas of knowledge and science? Is the market ready to receive knowledge-based products? Are there enough universities and research centers available for development and growth of the STP? (UNESCO, 2012).

Policies for Development of STPs

A number of policies can help foster the development of STPs, and these are particularly important for a developing country. Among the most important are:

First, they must have the ability to attract foreign and international ventures and companies. This can be accomplished by giving these businesses some privileges and preferential treatment—for example, special siting for new facilities, simplified registration procedures, extension of credit, and multi-entry visas for technical personnel.

Second, academic institutions and universities must attract talented researchers. The role of science and technology parks is to direct their skills toward economic and industrial growth.

Third, local government must encourage STPs so as to attract new companies to towns and to expand their tax base and employment opportunities to citizens. For example, property taxes and other taxes might be waived for a certain number of years, in order to attract new companies to the STPs.

Fourth, protection of intellectual property rights is essential if individuals and organizations will be contributing entrepreneurial skills and resources toward the development of new technologies.

Other policies might include special subventions for work permits for scientists and engineers returning from abroad, or for foreign technical specialists coming to work in the STPs. Additional forms of support, such as research funding, bonuses, or allowing workers to share the economic returns of their companies, may also be considered.

Models for STPs

The initiative for the development of STPs depends to a great extent on the country. In the United States, it usually comes from the universities, supported by state or local governments, with little intervention from the federal government. In France or Spain, development depends more on the central or regional government. A number of different modes exist.

US model

In the United States, STPs are typically initiated by universities, with state or local government support and negligible federal intervention. Stanford University's model (Silicon Valley STP) was the pioneer, followed by other parks, such as Boston's Route 128, the Seattle Excellence Area, the semiconductor industry in Minneapolis–St. Paul, Philadelphia, and Tucson, and the science parks of Triangle Park and Duke University in North Carolina.

UK model

Great Britain was the STP pioneer in Europe. Universities were responsible for creating the parks, as they had abundant real estate and experience in technology transfer. The parks are located on or near the universities' campuses. The first parks, created in the early 1970s, were the Cambridge Science Park, and the Heriot-Watt University Research Park in Edinburgh.

The British parks are different from the American ones in the size of the firms, as they tend to be smaller with fewer employees. Their main promoters are universities and local and regional authorities, some regional development agencies, and banks.

French model

In France, the cities and city councils use their own resources to create "poles of excellence" (APTE, 2012).

Sophia Antipolis, one of the famous parks in France, is a model for the "technopolitan" style. It is located on the Côte d'Azur as an attempt to modify the economic structure of the region, which had been based on tourism. The two most prevalent models in France are the "pole model" and the "agglomeration model." The "pole model" promotes the idea of concentrating a set of activities that did not previously exist in the region; an example of this is the Sophia Antipolis. The "agglomeration model," such as the ones in Montpellier and Lyon, encourages the global expansion of the influence of their location.

The French *technopoles* originated from a single pole (or park) in the 1980s. Over time, they developed a wider or multi-pole strategy. A good example is Montpellier Europole, which integrates five specialized poles: Euromédicine for healthcare; Agropolis, specializing in food and farming: Antenna, in multimedia; Communicatique, in computers and robotics; and Heliopolis, for tourism and leisure activities.

German model

Germany is known for its network model, whose base is technology transfer. The most important technology transfer pole in Germany is the Baden-Württemberg region, with over one hundred centers. Germany has the greatest number of parks in Europe, followed by Spain.

Examples of STPs

Japan

After its massive destruction in the Second World War, Japan worked hard to become one of the top five most technologically advanced countries in the world. Japan now has the three best science and technology parks: Yokosuka Research Park, Kansai Science City, and Tsukuba Science City.

Yokosuka Research Park (YRP), constructed in the 1990s, is located in Yokosuka City, next to NTT's Yokosuka Research and Development Center. The park hosts numerous wireless and mobile communications–related companies that have set up their research and development centers and joint testing facilities. This park has contributed to many communication breakthroughs that have been of global benefit. It was in Japan that 4G technology was first used (UNESCO, 2012).

Kansai Science City is another example in Japan of the private sector contributing to research in the domains of natural sciences, social sciences, and humanities. The city's facilities are distributed among twelve research districts, all connected by computer networks in order to facilitate collaboration among institutes in the city. Examples of research facilities for advanced science and technology are the national Nara Institute of Science and Technology, founded in 1991, the Advanced Telecommunication Research Institute International (ATR), founded in 1989, and the Research Institute of Innovative Technology for the Earth (RITE), founded in 1990 (IASP, 2006).

Tsukuba Science City is one of the best science cities in Japan; it has made global contributions in improving the quality of scientific discovery. Land was designated for this city in the 1960s. Construction began in the 1970s and it became operational in the 1980s. By the year 2000, it had sixty national research institutes and two universities grouped into five zones: (1) higher education and training, (2) construction research, (3) physical science and engineering research, (4) biological and agricultural research, and (5) common (public) facilities. More than 240 private research facilities surround these zones. It also includes the University of Tsukuba (formerly Tokyo University of Education); the High Energy Accelerator Research Organization; the Electrotechnical Laboratory; the Mechanical Engineering Laboratory; and the National Institute of Materials and Chemical Research.

Tsukuba Science City has an international flair, with about 3,000 foreign students and researchers from as many as 90 countries living there at any one time. Over the past several decades, nearly half of Japan's public research and development budget has been spent in Tsukuba (University of Tarbiat Modares, 2012). Important scientific breakthroughs by its researchers include the identification and specification of the molecular structure of superconducting materials, the development of organic optical films that alter their electrical conductivity in response to changing light, and the creation of extreme high-pressure vacuum chambers. Tsukuba has become one of the world's key sites for government–industry collaborations in basic research. Earthquake safety, environmental degradation, roadways, fermentation science, microbiology, and plant genetics are some of the broad research topics that benefit from close public–private partnerships here (Stone Creek Partners, 2012).

In today's global economy, it is vital for nations and regions to maintain their competitiveness. A major factor in this is the development of innovative products and services, which in turn depends significantly upon the efficient transfer of technology from academic and research institutions into industry. Many industrialized nations now speak of a 'knowledge-based economy.' Science and technology parks are one means of facilitating the growth of these industries (Wessner, 2009).

Hong Kong

The Hong Kong Science and Technology Parks Corporation (HKSTPC) is considered one of the leading parks in Asia. It was inaugurated on 7 May 2001. The Hong Kong government founded it as a non-profit statutory body. The HKSTPC is leading the transformation of Hong Kong into Asia's center for technology in many focused clusters, such as green technology and biotechnology. The program was first called "Incu-Tech" and had been operating since 1992. In the beginning it was located in a downtown building in Kowloon Tong. It was then moved to the Science Park in 2001, in a waterfront countryside area.

HKSTPC provides management and technical support to industry through collaboration with universities. It also offers technology startups through the Incu-Tech support program at a Tech Centre and provides advanced facilities and support services in the 22-hectare state-of-the-art Hong Kong Science Park for applied research and development activities. It also offers land and premises in its three Industrial Estates, totaling 239 hectares, for high-tech manufacturing. The project also provides advanced support services and facilities for high-technology companies, such as the IC Design/Development Support Centre and a photonics development support center, in addition to providing access to the best scientific and business minds in Hong Kong, China, and the world.

United Kingdom: Cambridge Science Park

Cambridge Science Park is one of the largest and oldest science parks in the United Kingdom. It is located on the northeastern edge of the city of Cambridge on land that had previously belonged to Trinity College since 1546. The idea started in 1970, when the government urged more interaction between UK universities and industries in order to improve technology transfer and increase the payback on investment in basic research and higher education. This expansion of "science-based industry" was recommended to be close to Cambridge, where there is a concentration of scientific expertise, equipment, and libraries, as well as a long tradition of scientific research and innovation dating from Sir Isaac Newton.

Planning permission was obtained in October 1971, and Laser-Scan, the first company, arrived in autumn 1973. Because Cambridge was the center of research in the United Kingdom, more companies were attracted. The Trinity Center—a meeting place, facilities, and conference rooms—was opened for the increasing number of people working in the park. In 1986 the Cambridge Innovation Centre and a squash court were built.

A number of venture and tenant companies found the park attractive, including 3i (the United Kingdom's leading venture capital company), Qudos (which was founded by the university's Microelectronics Laboratory), Prelude Technology Investments, and Cambridge Consultants.

As the number of high-tech companies increased to 1,200, employing around 35,000 people, more space was needed. At the beginning of the twenty-first century, a joint venture was launched between Trinity College and Trinity Hall (another Cambridge college). Neighboring land of about 22.5 acres owned by Trinity Hall was developed to accommodate five new buildings. In September 2000, Trinity Centre was opened, including a number of facilities such as a new conference center, broadband services, and a park-wide CCTV system. In 2005, the Cambridge Science Park Innovation Centre was established. Recently the park welcomed "Building One Zero One"—a 17-million-euro, 80,000-square-foot project—the famous software company Citrix, and a research and development building that has already attracted the Dutch electronics giant Philips.

United States

In 1963 in Pennsylvania, the University City Science Center in Philadelphia was established next to the University of Pennsylvania and Drexel University. This successful center was owned by a consortium of more than thirty academic and scientific institutions, hosting more than 200 technology and research-based organizations with 7,000 employees. The center is considered the world's first and most successful business incubator, as it has launched approximately 250 private-sector companies during the past three decades.

Another US park, the Research Park at Florida Atlantic University, is the only state-university-affiliated research park in south Florida. It hosts 22 high-tech research tenants and five support organizations. There are also 22 tenants in the Technology Business Incubator (TBI), which is managed by a very successful regional economic development engine, Enterprise Development Corporation of South Florida.

China

The Chinese government has relied heavily on science and technology parks as a cornerstone for industrial and technological development in China. This policy started in the late 1980s. The production of science and technology parks has grown at an average rate of 40% throughout the 1990s and until 2006 (Zhang and Sonobe, 2011). In fact, by 2006 Beijing had twice the number of firms participating in science and technology parks as in 2001—a total of 18,069.

In March 1991, the China State Council approved the establishment of 27 technology parks, followed by another 25 in the following year. The opening of the Yanglin Agricultural Technology Park in the western Chinese province of Shanxi in 1997 brought the total number of national technology parks to 53. In the meantime, a large number of technology parks have also been established by various levels of local government (Hu, 2007). These hundreds of STPs in China were considered the spark to ignite national development in China.

The National Science Council of the People's Republic of China believed that science and technology parks could serve as an incubating base for private-sector businesses, as a training base for individuals with new, innovating skills, and as an experimental zone for new technologies. They classified their STPs as follows: "26 per cent are involved in information and telecommunications technologies, 27 per cent in biotechnologies, 19 per cent in electronics and computers, 9 per cent in agro-food, 8 per cent in environmental technologies, 6 per cent in new materials, and 5 per cent in pharmaceuticals." They added that "in terms of tenants in the existing science and technology parks, 51 percent are services companies, 26 percent are research activities, and 23 percent are industrial companies" (Hu, 2007).

According to Zhang and Sonobe (2011), the formation of new science and technology parks is also a solution to the problem of congestion of resources that affects productivity in these nations. This is because the development of such centers causes a shift in the technological know-how of these firms; they expand very rapidly, dispersing the resources over a larger variety of projects and hence providing better resource allocation on the aggregate level.

The Zhongguancun Science and Technology Park was the first state-level high-technology development zone in China. During the last twenty years, Zhongguancun has gathered around 20,000 high-tech and new-tech enterprises, such as Lenovo and Baidu, and has formed a high-tech and new-tech industrial cluster featuring different fields: electronic information, biomedicine, energy and environmental protection, new materials, advanced manufacturing, aerospace, research and development, and service.

Zhongguancun is the most intensive scientific, education, and talent resource base in China. It contains about 40 colleges and universities, such as Beijing University and Tsinghua University, more than 200 national scientific institutions, such as the Chinese Academy of Sciences and the Chinese Academy of Engineering, 67 state-level laboratories, 27 national engineering research centers, 28 national engineering and technological research centers, 24 university science and technology parks, and 29 overseas student pioneer parks (www.chinadaily.com.cn/m/beijing/zhongguancun/).

India

In the early 1990s, the Indian government established a network of national software technology parks that provided broadband connectivity based on satellite and fiber technology, a single-window clearance system for software exporters, and incubation services. These research parks have helped India generate a substantial return on national investments in research and training in science and engineering. Today, the software industry is dominated by globally competitive champions from India, including Tata Consultancy Services (TCS), Infosys, and Wipro Technologies, each of which has generated revenues in excess of US$1 billion a year in recent times (National Academies Press, 2012).

Spain (APTE, 2012)

The first technology parks appeared in Spain in the mid-1980s, as some of the most industrialized autonomous communities saw in technology parks a tool for the modernization of the traditional industrial structure and for diversification into new sectors, which could contribute to regional economic development.

STPs in the Middle East

Egypt

In Egypt, the idea of science and technology parks started in 1993, about the same time as in China, but unfortunately, Egypt did not obtain the same results. In Egypt, there is no interaction between business and universities. There are many reasons behind this, but mainly it is a problem of trust. The quality of education, universities, and labs makes industries and executives mistrust research professors and their results. They feel that academic research is a risky investment; thus there is no link between universities and industry.

A few projects were initially planned to be technology parks, but were converted to other purposes. One example is a park located in Burj al-Arab, Alexandria, which was initially planned to be a technology park (Mubarak STP) but was converted to a research center, known as Mubarak Research Center.

Another example is Smart Village, which was first started as an STP for IT and then converted to real estate. Egypt has many research centers, but unfortunately still has no real STP. Recently, the Zewail City of Science and Technology project was begun. It would be the first STP in Egypt, and is expected to be finished in the next few years.

Mubarak City for Scientific Research and Technology Applications (MuCSAT)
According to the UN Industrial Development Organization (UNIDO, 2015), there are four science and technology parks in Egypt. The newest of these is Mubarak City for Scientific Research and Technology Applications (MuCSAT). It was established in 1993, occupies 101 hectares, and is managed by a board of trustees headed by the minister of higher education and scientific research. The park's main target sectors are biotechnology, information technology, advanced engineering, and nanotechnology. To support these sectors, the park will develop twelve research centers at various intervals.

Egypt's Smart Village (ESV)
Smart Village is Egypt's first initiative toward a high-tech environment. Sited on 121 hectares, it was built to attract major multinational and local companies in the fields of telecommunications and information technology. It boasts state-of-the-art infrastructure and will accommodate about 30,000 employees when it is completed (UNIDO, 2015).

Zewail City STP (ZCSTP, 2012)
Zewail City of Science and Technology was conceived by the Nobel Prize winner Dr. Ahmed Zewail in 1999. The government provided 121 hectares in Sixth of October City, and the cornerstone was laid on 1 January 2000. The project consists of three constituents—Zewail University, research institutes, and the technology park—and it is expected to be opened in the next few years as a center for science.

Zewail City of Science and Technology will be the first actual STP in Egypt. The project aims to raise local technologies to the world level and contribute to a knowledge-based society. Several such advances have been incorporated into its landscaping and facilities since January 2012, and other new facilities, including a number of research centers, a laboratory building, and a theater, are expected to be finished in the future.

The city is governed by a board of trustees, with Dr. Ahmed Zewail as the chair. The board includes six Nobel Prize winners in physics, chemistry, medicine, and economics; the current president of the US National Academy of Engineering (and former president of MIT); the current president of Caltech; the president of the Chinese Academy of Sciences; and world-renowned scientists, medical doctors, engineers, economists, and entrepreneurs.

The initial phase will include up to twelve institutes, as dictated by developments in the scientific landscape worldwide, but the criteria for selecting new institutes will be the quality of researchers and the funding available. The board of trustees approved the establishment of seven research centers.

Sinai Technology Valley (STV)
The Sinai Technology Valley, covering an area of 72 square kilometers, is located at the northwestern access to the Sinai Peninsula, on the east bank of Suez Canal, within the territorial jurisdiction of Ismailia Governorate. It is also called "Silicon Valley Egypt" because the Sinai Valley contains the highest-quality sand in the world. The park hosts enterprises in the sectors of microelectronics, biotechnology, new materials, fine tools, and renewable energy (UNIDO, 2015). They will be implemented in five stages. The investments in the first stage amounted to nearly LE 500 million.

Northern Coast Technology Valley (NCTV)
According to the UN Industrial Development Organization (UNIDO), this project is still under development. It is being studied by the Alexandria governorate, the Ministry of Higher Education, the Ministry of State for Scientific Research, and the Social Fund for Development.

Qatar (QSTP, 2011)
Qatar Science and Technology Park (QSTP) is the first STP in Qatar, located at Qatar Foundation's Education City, twenty kilometers from Doha International Airport. It was established in 2004 as a part of Qatar Foundation, with the aim of spurring development of Qatar's knowledge economy. It is a national project that aims to carry out applied research and deliver commercialized technologies in four areas: energy, environment, health sciences, and information and communication technologies.

QSTP hosts international technology companies in Qatar, and acts as an incubator of start-up technology businesses. It provides office and lab space to tenant companies, in a complex of multi-user and single-user buildings, as well as professional services and support programs. The science park became a 'free zone' in September 2005, allowing foreign companies to set up a 100-percent-owned, duty- and tax-free entity. Its location at Qatar Foundation's Education City, with leading international universities, allows its tenant companies to collaborate with the universities and act as an incubator for spin-out ventures from the universities (and other sources).

Saudi Arabia

The Kingdom of Saudi Arabia has recently engaged in a very ambitious industrial development plan, backed strongly by the political leadership, that depends heavily on the development of large-budget STPs. This is a growing trend in the Gulf countries in particular, and several Arab states are expected to engage in this trend in the near future (UNIDO, 2015).

The science parks founded in Saudi Arabia follow the classic example that depends on resourceful academic entities. The first example is the Prince Abdullah Bin Abdel Aziz Science Park (PASP). The park is closely linked to the King Fahd University for Petroleum and Minerals. It is sited on 300,000 square meters and is mainly concerned with oil, mining, and information technology applications. Among the notable contributors in this entity is Schlumberger and Jeddah Biocity Science Park, dedicated to biotechnology research (UNIDO, 2015).

Among other promising recent endeavors in Saudi Arabia is the King Abdul-Aziz University for Science and Technology (KAUST). This institution was founded as a massive incubator for researchers and individual entrepreneurs. The university is still expected to support large-scale corporations and firms with extensive industrial experience. However, transferring the concept of STPs to a wide range of clients is a unique initiative of the university. This step is expected to magnify the benefits of the integration that these parks provide so that these benefits reach small-scale firms as well as giant corporations.

STPs for Sustainable Development: Examples of STPs at Work

As previously defined, 'sustainable development' is "development that meets the needs of the present without compromising the ability of future generations to meet their own needs" (WCED, 1987). The key concepts in this definition are 'needs,' especially of the world's poor, which are of high priority, and 'limitations of natural resources' arising from technological and social pressures that hinder the environment's ability to meet present and future needs.

The major economies in the twenty-first century are knowledge-based economies characterized by high levels of skills and education, lifelong learning, and innovation in all areas of industrial and commercial competitiveness. Knowledge-based economics is based on a brain-driven system, not a money-driven system as in the past. Brain-driven development requires modern tools and technology, and gives high priority to science, technology, and innovation. This is why STPs are so important for promoting sustainable development and a sustainable economy. Fostering innovation requires attention to all components of the innovation ecosystem. It must include a comprehensive national innovation strategy that can stimulate dialogue between public and private bodies, and between university and research institutes, about specific measures to take full advantage of the national potential for innovation that results in economic growth.

Innovative STPs work with different organizations in society, such as NGOs and public and private organizations, to create awareness about innovation and sustainable development and to engage professors and graduates to find ways of sustaining natural resources for future generations. These professors and graduates come not only from science and engineering specializations, but also from the humanities, business, economics, marketing, and public affairs, with the goal of inventing easy and inexpensive technologies to utilize pollution as a natural resource. They provide products and services, and cooperate with society to create new job opportunities and to market new products within and outside the country.

The new development of science and technology parks for sustainable development (STPSD) can be considered as the main tool for starting the next industrial revolution, according to C2C thinking, and for implementing a sustainable, green economy, as explained in chapter 3. It will have a significant impact not only on the national economy but also internationally.

Based on the idea of a cradle-to-cradle (C2C), sustainable economy, sustainable development can be redefined as **development that meets the needs of the people according to C2C concepts in order to attain a sustainable economy and move beyond sustainability**.

An STPSD consists of a number of industrial clusters/complexes (incubators). Each of these clusters is looking for a solution to one of the national sustainable-development problems. The clusters overlap: wastes from one cluster may be mixed with another type of waste from another cluster to produce high-quality products (upcycling) that meet international standards. Some of these types of clusters are: agricultural waste; municipal solid waste; marble and granite waste; industrial waste; sustainable food–feed–fuel; construction and demolition waste; energy saving and sustainable energy; water saving and sustainable waste-water treatment; and many others. Each of these clusters is being addressed by STPSDs in different parts of the world, in search of sustainable and economically viable solutions to these problems.

Summary

As the world continues to produce technological innovations, all countries should be aware of this transition and should prepare to compete in this new era. This chapter discusses the main features of science and technology parks (STPs) in general, and science and technology parks for sustainable development (STPSDs) in particular. An STP is a cluster of technology-based organizations situated on or near a university campus in order to benefit from the university's knowledge base and ongoing research, and to develop commercial applications for this knowledge in association with the commercial tenants in the park. STPSDs consist of a number of industrial clusters/complexes (incubators), where each cluster/complex will devote itself to a specific national problem according to the standards of sustainable development/sustainable economy.

Based on the ideas of cradle-to-cradle (C2C), sustainable economy, 'next industrial revolution,' STPSDs, and so on, sustainable development can be defined as development that meets the needs of the people according to the C2C concept in order to reach a sustainable economy and to advance beyond sustainability. The STP is considered the first step toward the next industrial revolution.

STPs can play a significant role in the development of nations. They contribute to economic growth and strength and share resources, such as a continuous power supply, centers of communication, banks, restaurants, transportation, entertainment, and sports facilities.

All developed countries now have STPs, which have become public policy tools to promote local economic development and technological progress. STPs integrate scientific, technical, and social capabilities, facilitating the creation, transfer, dissemination, measurement, and management of knowledge, and apply them to production activities. They also strengthen the links between the firms located in the parks and their environment. They promote the productive use of knowledge and benefit their respective nations by disseminating their innovations.

Questions
1. What are the challenges facing STPs?
2. Discuss the main objectives for establishing STPs.
3. What is the difference between STPs for sustainable development and traditional STPs?

4. What is the role of innovation in STPs?
5. Discuss the main stakeholders involved in establishing a successful STP, and the role of each.
6. Discuss the advantages of connecting a university with an STP. Can an STP offer a technical scientific degree?
7. What do we mean by entrepreneurs? Explain the mechanisms for the development of entrepreneurship.
8. What are the necessary policies for STP development?
9. Discuss different models used for establishing STPs worldwide. Select the best model for your country and explain why.
10. Explain the link between green economy and STPs and/or STPSDs, with examples.

References

Alberto, A., S. Perez-Canto, and P. Landoni. 2010. "Science and Technology Parks' Impacts on Tenant Organizations: A Review of Literature." mpra.ub.uni-muenchen.de/41914/1/MPRA_paper_41914.pdf

APTE (Asociación de Parques Científicos y Tecnológicos de España). 2012. "A Study of the Social and Economic Impact of Spanish Science and Technology Parks." www.apte.org/en/documents/estudiodeimpacto_ingles.pdf

AURP (Association of Universities and Research Parks). 2013. "What Is a Research Park?" www.aurp.net/what-is-a-research-park

Hewes, A., and D. Lyons. 2008. "The Humanistic Side of Eco-Industrial Parks: Champions and the Role of Trust." *Journal of Regional Studies* 42, no. 10: 1329–42. Business Source Complete. www.techpark.ir/files/pdf/statistics.pdf

Hu, A.G. 2007. "Technology Parks and Regional Economic Growth in China." *Research Policy* 36: 76–87. www.sciencedirect.com

IASP (International Association of Science Parks). 2015. "About Science and Technology Parks: Statistics." International Association of Science Parks. 25 February. www.iasp.ws/statistics

Monck, C.S.P., R.B. Porter, P. Quintas, D.J. Storey, and P. Wynarczyk. 1988. *Science Parks and the Growth of High Technology Firms*. London: Croom Helm.

Moudi, M., and H. Hajihosseini. 2011. "Science and Technology Parks: Tools for a Leap into Future." *Interdisciplinary Journal of Contemporary Research in Business* 3, no. 8: 1168.

MuCSAT. 2011. "Technology Parks in Egypt: Mubarak City for Scientific Research and Technology Applications (MUCSAT)." www.unido.org/index.php?id=o25786

National Academies Press. 2012. "Understanding Research, Science and Technology Parks: Global Best Practice: Report of a Symposium." www.nap.edu/openbook.php?record_id=12546

QF (Qatar Foundation). 2011. "Qatar Foundation." www.qf.org.qa/science-research/science-research-institutions/qatar-science-technology-park

QSTP (Qatar Science and Technology Park). 2011. "Qatar Science and Technology Park." www.qstp.org.qa/output/page7.asp

So, S. 2011. "Incubation Program at Hong Kong Science and Technology Park." TechNode, 6 September. technode.com/2011/09/16/incubation-program-hong-kong-science-and-technology-park/

Stone Creek Partners. 2012. "Science and Technology Parks." Stone Creek Partners Resource Page. www.stonecreekllc.com/science-technology-parks.html

"Survey of Energy Resources." 2001. World Energy Council. www.worldenergy.org/wecgeis/publications/reports/ser/coal/coal.asp

SVE (Smart Village Egypt). 2011. "Smart Village Egypt." www.smart-villages.com/docs/cairo.aspx

UKSPA (United Kingdom Science Parks Association). 2013. "International–Overseas Application Form." www.ukspa.org.uk

UNESCO. 2012. "Science Parks around the World." *Science Policy and Capacity-building.* www.unesco.org/new/en/natural-sciences/science-technology/university-industry-partnerships/science-parks-around-the-world/

UNIDO. 2015. http://www.unido.org/en/how-we-work/convening-partnerships-and-networks/networks-centres-forums-and-platforms/technology-parks/mapping/mena-region/egypt.html

University of Cambridge. www.cambridgesciencepark.co.uk/about/history/

University of Pune. 2012. Scitech Park. www.scitechpark.org.in/index.php/component/content/article/44-home

University of Tarbiat Modares. 2012. "Science and Technology Town." www.modares.ac.ir/en/Administration/aft/SST

WCED (World Commission on Environment and Development). 1987. *Our Common Future.* Oxford: Oxford University Press.

Wessner, C.W. 2009. *Understanding Research, Science and Technology Parks: Global Best Practices.* Washington, DC: The National Academies Press. http://www.nap.edu/openbook.php?record_id=12546

ZCSTP (Zewail City Science and Technology Park). 2012. www.zewailcity.edu.eg

Zhang, H., and T. Sonobe. 2011. "Development of Science and Technology Parks in China, 1988–2008." *Economics: The Open-Access, Open-Assessment E-Journal* 5:2011–16.